Multidisciplinary Research for the Monitoring and Preventive Conservation of Cultural Heritage

Multidisciplinary Research for the Monitoring and Preventive Conservation of Cultural Heritage

Guest Editors

Silvana Fais
Giuseppe Casula

MDPI

Basel • Beijing • Wuhan • Barcelona • Belgrade • Novi Sad • Cluj • Manchester

Guest Editors
Silvana Fais
Department of
Civil—Environmental
Engineering and Architecture
University of Cagliari
Cagliari
Italy

Giuseppe Casula
Istituto Nazionale di Geofisica e
Vulcanologia (INGV)
Bologna
Italy

Editorial Office
MDPI AG
Grosspeteranlage 5
4052 Basel, Switzerland

This is a reprint of the Special Issue, published open access by the journal *Minerals* (ISSN 2075-163X), freely accessible at: https://www.mdpi.com/journal/minerals/special_issues/9OM513G8K4.

For citation purposes, cite each article independently as indicated on the article page online and as indicated below:

Lastname, A.A.; Lastname, B.B. Article Title. *Journal Name* **Year**, *Volume Number*, Page Range.

ISBN 978-3-7258-5383-0 (Hbk)
ISBN 978-3-7258-5384-7 (PDF)
https://doi.org/10.3390/books978-3-7258-5384-7

Contents

About the Editors

Silvana Fais

Silvana Fais is associate professor at Cagliari University (Italy). Her research is focused on the application of geophysical methods to mining and environmental problems, as well as on the integrated application of non-invasive techniques applied to the diagnostics of cultural heritage. She founded the Laboratory of Solid Earth Physics and Diagnostics at the Department of Civil, Environmental Engineering and Architecture of the University of Cagliari. Throughout her career, she has coordinated approximately 50 research projects funded by various public and private institutions. She has supervised nearly fifty graduate students at both master's and doctoral levels, published numerous scientific articles in peer-reviewed journals, and presented over 80 abstracts at international conferences. Additionally, she actively serves as a reviewer for international journals and as an evaluator for national and international research projects. She is also a member of research task teams at the Institute of Environmental Geology and Geoengineering (IGAG) of the National Research Council (CNR) of Italy and the National Interuniversity Consortium for Georesource Engineering (CINIGEO) in Rome (Italy).

Giuseppe Casula

Giuseppe Casula holds a degree in Applied Physics with a specialization in Geophysics from the University of Bologna (Italy), obtained in July 1990, discussing a thesis on the use of LaCoste-Romberg gravimeters in geophysics. He currently holds a permanent position as senior technologist at the Bologna Unit of the National Institute of Geophysics and Volcanology (INGV). He is a geodesist specialized in Gravimetry, Global Navigation Satellite Systems (GNSS), satellite remote sensing technologies, photogrammetry, and terrestrial LiDAR, with applications in 3D modelling and the study of cultural heritage. Dr. Casula has authored over 160 publications in international journals, many of which are indexed in major databases such as Google Scholar, Scopus, and Web of Science. He has contributed to several research projects (Global Geodynamics Project (GGP), PON Massimo) and serves as an editor and reviewer for some international Geophysical journals.

Preface

Currently, the available technologies and methodologies for the monitoring and preventive conservation of cultural heritage are promising, and the scientific community has been finalizing studies to find approaches fast with effective and easy 3D documentation. A significant part of the recent scientific research in various fields of applied sciences is focused on the integration of datasets of different natures and complementary methodologies aimed at solving specific problems and reducing uncertainties. In the cultural heritage field, the integration of different types of complementary information can greatly improve the diagnostic process for the conservation state of monuments and artworks. The diagnostic process of stone material decay is very complex and cannot be described by a single discipline. The prevention and rehabilitation of monumental structures or works of art in a visual form, such as paintings or sculptures, can only be successful by combining different methodologies. The multidisciplinary approach, starting from historic and architectural analysis to a complete mechanical, physical, mineralogical, chemical, and petrographic characterization of the different materials by optical and electronic microscopy, is believed to have the greatest chance of success. In fact, in complex situations, the use of a single methodology is insufficient to obtain a reliable diagnostic process. The integration of different datasets is a useful process that helps the decision making during the restoration actions.

In synthesis, the need to integrate different methodologies in the diagnostics of cultural heritage derives from two main assumptions: first, the complexity of the factors that contribute to the degradation of monuments and works of art in which the diagnosis would be very difficult using a single methodology. Second, the technological development and innovation in applied scientific research that allow us to acquire and process large multidisciplinary datasets.

Moreover, monumental structures and heritage sites are at risk from pollution, tourism, and geo-environmental disasters such as earthquakes and climate change. Their protection is the duty of future generations.

In this framework, the Special Issue covers several research and aims to improve knowledge and plan restoration solutions. Therefore, the main target of this special volume is to highlight the need to integrate different disciplines and different professional skills during the process of diagnosing materials of a different nature that constitute monumental heritage and works of art. Consequently, this reprint is aimed at a large audience of applied scientists, such as architects, restorers, geologists, applied petrographers, and civil engineers, who consider the case studies presented in this reprint to be of their interest.

This Special Issue features eight articles, four of which focus on the characterization of stone materials through both in situ and laboratory studies utilizing non-invasive diagnostic techniques. These studies are further complemented by petrographic analyses and geological knowledge. The results demonstrated that integrating various non-destructive techniques (e.g., geomatical and geophysical) allows for an accurate assessment of the conservation status and intrinsic characteristics of stone materials used in building monuments. This Special Issue also includes four articles focused on the use of non-destructive techniques for the detailed investigation and analysis of paintings by various artists, as well as mural paintings. The non-invasive techniques applied by the authors really helped uncover valuable information about the materials and techniques used by painters, which are very helpful for both art conservation and historical research.

Silvana Fais and Giuseppe Casula

Guest Editors

minerals

MDPI

Article

Material and Technical Analysis of La Inmaculada by Francisco Pacheco

Anabelle Kriznar [1], Javier Moreno-Soto [2,3], Antonio Gamero-Osuna [4], Agustín Martín-de-Soto [4], Francisco José Ager [3,5] and Miguel Ángel Respaldiza [3,6,*]

1 Departmento de Escultura e Historia de las Artes Plásticas, Facultad de Bellas Artes, Universidad de Sevilla, C/Laraña 3, 41003 Seville, Spain
2 Departamento de Física Aplicada I, Escuela Técnica Superior de Ingeniería Informática, Universidad de Sevilla, Avda. Reina Mercedes s/n, 41012 Seville, Spain
3 Centro Nacional de Aceleradores (US), Avda. Thomas A. Edison 7, 41092 Seville, Spain; fjager@us.es
4 Conservation-Restoration Workshop, Archbishop's Palace, Plaza Virgen de los Reyes, 41001 Seville, Spain
5 Departamento de Física Aplicada I, Escuela Politécnica Superior, Universidad de Sevilla, C/Virgen de África 7, 41011 Seville, Spain
6 Departmento de Física Atómica, Molecular y Nuclear, Facultad de Física, Universidad de Sevilla, Avda. Reina Mercedes s/n, 41012 Seville, Spain
* Correspondence: respaldiza@us.es

Abstract: Francisco Pacheco (1564–1644) was one of the most important Spanish painters, the author of the famous treatise Art of Painting, and a founder of Seville art school. His painting La Inmaculada (1610) forms part of the important art collection in the Archbishop's Palace in Seville and represents the first realization of a new iconography, established by Pacheco. Later, he carried out several paintings with the same subject which are today in different collections. As part of a larger project, La Inmaculada was recently restored and analysed. The characterization of materials and painting procedures was carried out applying noninvasive in situ analysis by ultraviolet fluorescence (UVF), infrared reflectography (IRR), and X-ray fluorescence spectrometry (XRF). Results revealed many retouches, probably from different periods, as well as some pentimenti in the composition. Chemical analysis indicates a preparation made of Seville clay, lead white and some calcite on an animal glue layer, according to Pacheco's own treatise, and a corresponding pigment palette for the 17th century: lead white, yellow and red earths, lead–tin yellow, vermilion, azurite, smalt, a copper-based green, umber, and bone black. Furthermore, modern pigments such as titanium and zinc whites, chrome green, and cadmium yellow were also identified as results of later interventions.

Keywords: Francisco Pacheco; La Inmaculada; supports; pigments; preparatory drawing; noninvasive analysis; UVF; IRR; XRF

1. Introduction

Francisco Pacheco (1564–1644) was already a renowned artist during his life and is still considered one of the most important Spanish painters of the Mannerist style [1,2]. He is well known for his art, as a teacher of Diego Velazquez and Alonso Cano, as well as for his treatise Art of Painting (Arte de la Pintura; [3]), published posthumously in 1649. It is an important source for studying the 17th century painting technique in Spain and a guide for many contemporary and later artists, since it precisely describes the materials (supports, preparation, pigments, binders) and the manner of painting in order to achieve desired results. In Seville, he founded an art school, where he emphasized the academically correct representation of religious iconography. One of his works represents the Virgin Mary, La Inmaculada (Figure 1), painted on canvas in oil technique around 1610–1615. It forms part of the art collections held at the Archbishop's Palace in Seville, one of the most prominent painting compendiums in Spain. It is considered the earliest and the simplest 17th century representation of a new iconography of the Immaculate Conception, depicted for the first time in this manner precisely by Pacheco. La Inmaculada Concepción de la Virgen (Immaculate Conception of the Virgin Mary) is one of the most repeated iconographies in the Spanish art since the 15th century and is very popular especially in Spanish Baroque art. The iconography is based on the Catholic dogma stating that the Virgin Mary's conception is free of the original sin. She is shown as a young woman, dressed in white tunic and blue coat, with symbols of purity and eternity, crowned by a nimbus of twelve stars, standing on top of a half-moon, sometimes with a snake at her feet (not in this Pacheco version), symbolizing her domain over the sin. Under her feet, a landscape with trees, buildings, sea, and a boat can be depicted, indicating the Virgin's role as the intermediary between the Earthly world and God. In this Pacheco version, the landscape shows contemporary Seville with its Torre del Oro. In later variants, the Virgin is usually accompanied by angels. Traditionally, the Virgin Mary is dressed in a red (not white) tunic which can symbolize her royalty or the Passion of Christ and can be still found in later Inmaculadas; however, the new white-and-blue vestment was painted for the first time in the Inmaculada studied in this research. This iconographic representation influenced many other Spanish painters such as Velázquez, Herrera, El Greco, Zurbarán, Ribera, Valdés Leal, and Murillo, while it also served as inspiration for Virgin's imagery in the holy processions in Seville [3–5].

Figure 1. Francisco Pacheco: La Inmaculada, 1610, Archbishop's Palace, Seville.

Later on, Pacheco carried out other Inmaculadas, several of which are still in Seville. Surprisingly, the artist decided to use the traditional red–blue vestment iconography. Among these are La Inmaculada con Miguel Cid (1619) in the Cathedral of Seville, La Inmaculada con Vázquez de Leca (1621) in a private collection of Marqués de la Reunion de Nueva España, and La Inmaculada from the church of San Lorenzo in Seville (1624). None of them have been studied from a materials point of view, but this is one of the research goals in the future, in order to compare them to the panel presented in this work.

During the last couple of years, the Archbishop's Palace patrimony has been systematically studied, restored, and conserved as a principal aim of a larger project. The present painting was one of the selected ones to be restored and studied; therefore, a concise material and technical research was carried out. The analytical study of this canvas was a once-in-a-lifetime opportunity, since it is part of the Archbishop's personal collection, where it was returned after its restoration. All available noninvasive analytical techniques at the time were used. Considering other Pacheco paintings with the same iconography, this study is a perfect start for any future analysis and technical–material comparison among them.

The painting went through an analytical process, in order to identify the support, original pigments, later interventions, possible preparatory drawings, and the painting technique in general. Obtained data offer a deeper insight into the understanding of Pacheco's way of painting and a more thorough comparison to other known works attributed to the artist, especially to other existent Inmaculadas. One of the most important objectives was to use this material analysis as support in the conservation–restoration process.

2. Materials and Methods

As it was not possible to extract samples, only noninvasive techniques available at that time were used: ultraviolet fluorescence photography, infrared reflectography (IRR), and X-ray fluorescence spectrometry (XRF). They were all applied in situ in the conservation–restoration workshop of the palace (Figure 2).

Figure 2. XRF in situ analysis of La Inmaculada.

2.1. Ultraviolet Fluorescence (UVF)

This technique [6–10] is generally used to identify later interventions, which greatly helps in the cleansing process of the painting surface. It is also a very useful tool for selecting original areas for the following chemical analysis. UVF images were taken with a Nikon D3X camera (Nikon, Tokyo, Japan), at 400 ISO and 20 s of exposure time, with the painting being homogeneously illuminated with four wood lamps, two on each side of the canvas, using the UVGL-55 model with double band wavelength (254 nm/365 nm), in a completely darkened room.

2.2. Infrared Reflectography (IRR)

IRR [6,7,11–13] can reveal possible underdrawings and *pentimenti*, changes in the composition. It can help with the authentication of a painting, generally revealing the proper handstroke of the artist. For this technique, a small 5 cm^3 XenICs (Leuven, Belgium) near-IR camera Xeva-XS 512 with InGaAs detector and Pentax lens of 16 mm, F/1.4 (320 $\times$ 256 pixel resolution) in the range of 900–1700 nm was used. The painting surface was illuminated with two 800 W halogen SDI lamps. The camera was mounted on a 2D robotic Optimind platform at 32 cm from the painting to obtain high movement precision along the studied surface.

2.3. X-ray Fluorescence Spectrometry (XRF)

XRF [6,7,14–17] can identify inorganic materials used in the artwork: preparation and priming, as well as original pigments and those applied in later interventions. It allows an elemental analysis, followed by the characterization of materials applied in the selected area, based on their characteristic chemical elements. Our XRF equipment was composed of a RX38 X-ray generator from EIS S.L. with W anode, 1 mm Al filter to suppress W peaks, and a Si drift detector (SDD). The measuring distance was fixed to 1 cm from the painting surface by the intersection of two lasers. The generator was always operating under the same conditions: 33.4 kV and 80 μA with 200 s acquisition time. The appearance of Zr peaks in every spectrum is due to its presence in the collimator of the detector. The areas of the XRF peaks observed in the spectra were used to perform a semiquantitative analysis of the colors and estimate whether a particular element belongs to the surface layer or to an inner layer of the painting.

3. Results

Results obtained by all selected analytical methods are presented and compared with available information found about Francisco Pacheco and materials he used to apply.

3.1. UVF and IRR Examination

The ultraviolet fluorescence revealed many smaller and larger retouches spread around the surface (Figure 3). They must have been carried out at different times, showing varied tonalities of material fluorescence. The largest interventions can be found on the left margin, next to the Virgin's right hip and left foot, revealing more important damages suffered by the painting. IRR confirmed several retouches and the use of different pigments as well; however, at first sight, no preparatory drawing can be appreciated (Figure 4). At a closer look, an underdrawing can be observed on the half-moon and on the boat (Figure 5a,b), while there is a small pentimento, a slight change in the composition of the architecture in the left bottom corner; the painter shortened some of its walls, painted with a reddish-brown color, obtained by mixing earths (Fe), a copper-based green pigment (Cu), lead white (Pb), and a small amount of umber (Mn, Fe) and vermilion (Hg), as studied later by XRF. An interesting observation is that the trees in the lower right corner almost disappear on the IRR image.

Figure 3. UVF image of La Inmaculada, showing many retouched areas, the largest one located next to the Virgin's left foot.

Figure 4. IRR image of La Inmaculada, confirming several retouched areas, the use of different pigments, underdrawing on the half-moon, and the change in the form of the architecture on the right.

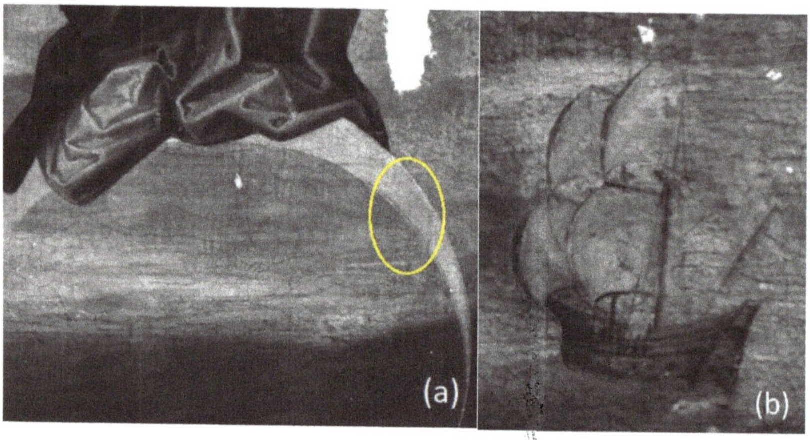

Figure 5. Detail of the IRR image, showing the underdrawing (**a**) on the half-moon and (**b**) for the design of the boat.

3.2. XRF Analysis of Materials Applied

Altogether, 38 points of different colors and tonalities were selected, in order to obtain knowledge not only on preparation and pigments, but also on their mixture and the construction of lights and shades. XRF is a great tool for the identification of most preparations and pigments; however, there are some withdrawals that call for the implementation of other complementary analytical techniques, which was not possible in this study. In this regard, pigments characterized by the same chemical element, such as copper or lead-based pigments, cannot be differentiated except by their visual color. On the other hand, organic materials such as lakes or binders that are composed of light elements cannot be identified by XRF, since it generally detects only elements heavier than $Z < 14$ (Si) [14,15,17].

3.2.1. Original Materials

Materials present in the painting were characterized: preparation, priming, and pigments. The summary is presented in Table 1. The presence of calcium (Ca) in all analysed areas suggests the use of calcium-based preparation, chalk, or gypsum. Nevertheless, from Pacheco's famous treatise Art of Painting [3] and some studies carried out on Andalusian contemporary painters [18–21], we know that Pacheco did not use gypsum for preparation layers (aparejo, as he calls it), finding it unstable as a painting support; the experience showed him that it moistens and rots with time, also affecting the canvas and color layers. Instead, he preferred well-grounded special Seville clay, tempered with linseed oil, to which he could also add lead white, as he describes in his text [3,20,22]. This explains the presence of Fe and Pb in all analysed areas, although both elements also identify related pigments applied as color layers, while lead compounds can be used as siccative, as well. From Pacheco's treatise we can also learn that before this earth-based preparation [preparación], he used to apply several layers of animal glue ("la cola de guantes flaca") over the entire surface of the prepared canvas to seal it. On the other hand, in other contemporary preparation layers, earths and clays could be combined with some pulverized calcium carbonate or some other pigments, as in the case of Murillo's primers [18,19,21]. Therefore, some chalk was probably also added to the La Inmaculada primer, which can explain the presence of Ca in all analysed points. This hypothesis can be supported by some studies carried out in the Museo de Bellas Artes de Sevilla on two Pacheco's paintings (St. Francis with the Cross and The Mystical Nuptials of St. Agnes), which confirm the reddish preparation made of earths and calcite, while in the case of St. Francis, also includes minium [20].

Table 1. Systematic representation of original and modern materials identified in La Inmaculada by XRF, with characteristic chemical elements for each material in bold.

Preparation, Pigments	Chemical Formula/Elements	Principal Use
	ORIGINAL MATERIALS	
Animal glue	Ca	Sealing layers
Seville clay + lead compounds	Fe, Pb	Preparation/priming
Lead white	$\mathbf{2PbCO_3}$·Pb(OH)$_2$	Flesh tones, highlights, clouds, sky, aureola, moon
Yellow earth	**Fe**(OH)$_2$	Aureole, crown, hair, clouds, town, landscape
Lead–tin yellow	$\mathbf{Pb_2SnO_4}$	Highlights—crown, trees, clouds
Red earth	$\mathbf{Fe_2O_3}$	Addition to other pigments
Vermilion	**HgS**	Flesh tones, lips, red and pink flowers
Red lake	Unidentified, **Ca** substrate	Pink flowers
Smalt	Diverse chemical compounds, containing **Co**, **Ni**, **As**, Ni, Si, K, Bi.	Sky, clouds, see, town, landscape
Azurite	2CuCO·Cu(OH)$_2$	Virgin's coat; in green hues?
Green—questionable use	A **Cu**-based compound	Landscape, flesh-tones?
Umber	$\mathbf{Fe_2O_3}$ + $\mathbf{MnO_2}$ + nH$_2$O + Si + Al$_2$O$_3$	Virgin's hair, ground, fountain, architecture, shades
Bone black	$\mathbf{Ca_3(PO_4)_2}$ + CaCO$_3$ + C	Shades, darkest hues
	INTERVENTIONS	
Chalk or gypsum	$\mathbf{CaCO_3}$ or $\mathbf{CaSO_4}$	Filler
Titanium white	$\mathbf{TiO_2}$	Retouches
Zinc white	**ZnO**	
Cadmium yellow	**CdS**	
Chrome green	$\mathbf{Cr_2O_3}$	

Regarding the original pigments that compose Pacheco's palette, they are all common for the 17th century [3,18,23–25]: lead white (Pb), yellow and red earths (Fe), lead–tin yellow (Pb, Sn), vermilion (Hg), smalt (K, Co, Ni, As), and azurite (Cu). A copper-based green pigment (Cu), umber (Mn, Fe), bone black (Ca), and an organic red lake on the calcium substrate are also probable. Lead white, identified by intense Pb peaks, was used for all white areas, such as the half-moon, white aura around the Virgin, for her white tunic, and was largely added for lighting other colors. It was also used as a principal pigment for flesh tones, mixed with other pigments such as earths (Fe), vermilion (Hg), and azurite or a green copper-based green pigment (Cu), depending on the lighter or darker flesh tonality that Pacheco wished to obtain. Vermilion is highly present in reddish areas such as the Virgins' cheeks and decreases in lightest flesh tones, where lead white predominates. Shades were carried out with higher presence of a copper-based pigment, some umber (Mg, Fe), and bone black (Ca), while in light tones they almost disappear (Figure 6).

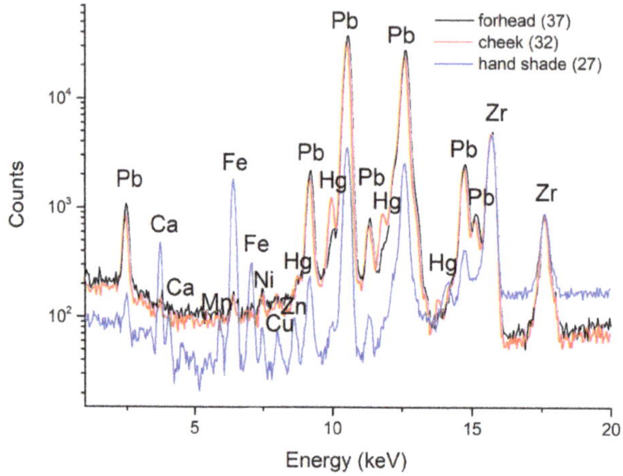

Figure 6. Comparison of XRF spectra of different flesh tones: light forehead (black line) with predominant presence of lead white (Pb), reddish cheek (red line) with higher amount of vermilion (Hg), and dark hand shade (blue line) with umber (Mn, Fe) and bone black (Ca).

The principal yellow pigment used was yellow ochre (Fe) that was applied for the Virgin's aureole, for her crown, and her hair. In low amounts it was added to other colors such as the sky, the landscape, or the sea below. However, for the highlights, Pacheco choose lead–tin yellow (Pb, Sn), which was identified on the crown (Figure 7), some trees, and a yellowish cloud. The principal red pigment was vermilion (Hg), used in flesh tones as well as, in high concentration, for the Virgin's lips and the red flower on her crown. Pink flowers were made with lead white and very little of vermilion; probably, a red lake was also used. Since the latter one is an organic colorant, it cannot be detected by XRF, but high Ca peaks in this area suggest a possible calcium substrate for the liquid lake. A small amount of vermilion was added to the brownish color of the architecture underneath and even for the shades of the blue coat and the tunic. Red earth (Fe) was probably used as addition to other pigments, as well.

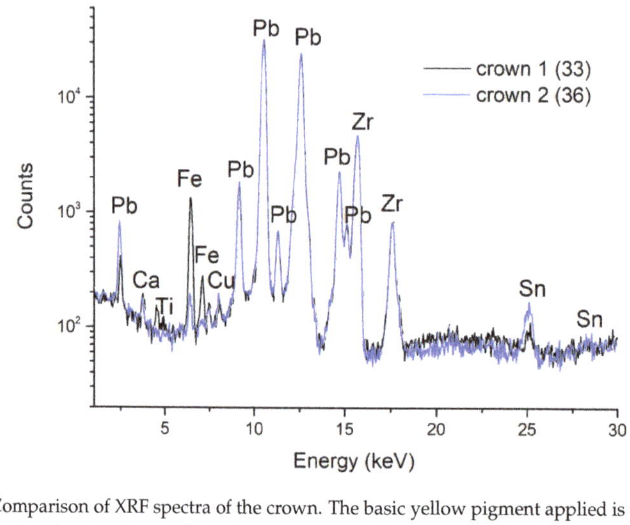

Figure 7. Comparison of XRF spectra of the crown. The basic yellow pigment applied is yellow earth (Fe) observed with black line, while highlights are carried out with lead–tin yellow (Pb, Sn) shown with blue line.

Large blue and greenish areas, such as the sky, the sea, the town on the horizon, and the landscape, were carried out with blue smalt (Figure 8), which in some areas lost its blue intensity due to chemical changes in the pigment, probably caused by the reaction with the oil binder or the instability of potassium glass [25,26].

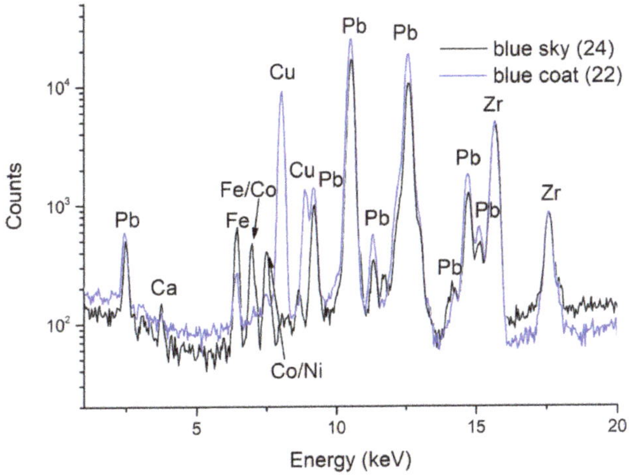

Figure 8. Comparison of two XRF spectra from blue areas reveal the use of azurite (Cu) for the Virgin's coat (blue line), and smalt (K, Co, Ni, As) for most of the background (black line).

Smalt is identified by common peaks of K, Ni, Co, and As; the high amount of Ni is surprising, its peak intensity being almost as high as that of Co, which was not detected in any of our previous painting studies. It must be a smalt compound due to a specific pigment preparation. On the other hand, a copper-based blue, probably azurite (Cu), an expensive pigment obtained from the semiprecious mineral, was selected for the Virgin's blue coat (Figure 8). This pigment is generally used for vestments of the most important sacred figures in a composition, revealing their celestial meaning. The blue coat of the Virgin symbolizes the eternity, as explained in the Introduction [4,5]. Azurite was also applied for the Virgin's blue eyes. Perhaps some azurite was added to smalt for the color of the sea; however, it could also be obtained as a mixture of smalt and a copper-based green pigment, Cu being the characteristic element in this bluish-greenish area. With XRF it is not possible to identify the green pigment precisely, but there exist several copper-based ones in the 17th century, such as malachite, verdigris, or copper resinate [14,18,24–26]. We can exclude verdigris, as Pacheco's own testimony in his treatise reveals that he did not use verdigris: "Keep away from verdigris like the plague, because it's your worst enemy" [3,22]. In La Inmaculada, a green color can be found in the landscape, mixed with a small amount of smalt. Darker tones and shades were modeled with umber (Mn, Fe), identified in the Virgin's hair basic color, the ground, the fountain, the architecture in the right corner, and on dark tones of the drapery and flesh tones. In the darkest areas and shades, high Ca peaks are observed, indicating a probable use of bone black.

3.2.2. Interventions

Following the UVF and IRR images, different retouches were confirmed by XRF as well, carried out by modern pigments such as zinc (Zn) and titanium (Ti) whites, chrome green (Cr), and cadmium yellow (Cd). In most cases, the white areas of zinc and titanium do not concur, indicating at least two different interventions, bearing in mind that the first pigment was in use since the second half of the 19th century, while the second one entered only in the 1920s [23–26]. The largest intervention is located in the lower part of the aureole, next to the Virgin's left leg (Figure 9). It is made on a calcium-based filler (gypsum or chalk), with both modern whites and a low amount of cadmium yellow (Cd) and chrome green (Cr), which was also located in the Virgin's left eye.

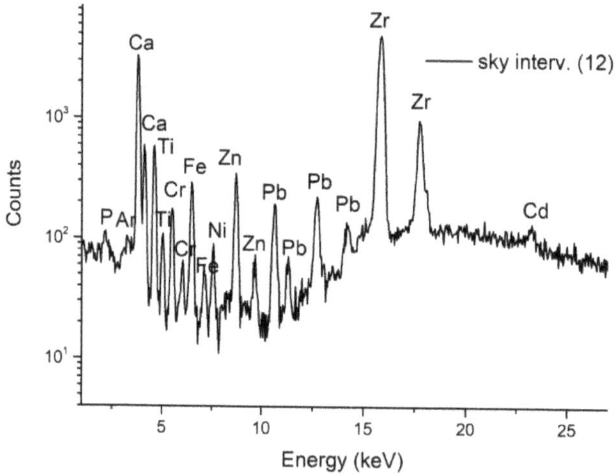

Figure 9. XRF spectrum of a retouched area in the sky, revealing the use of a calcium-based filler (probably gypsum), overpainted with modern pigments, zinc (Zn) and titanium (Ti) whites, chrome green (Cr), and cadmium yellow (Cd).

4. Discussion

The results obtained by these noninvasive analyses correspond to what it is known so far about Pacheco's painting technique from his treatise and from the analysis of his artworks performed in other institutions [3,18–21]. Even if the artist preferred to use wooden panels, especially oak when possible, this painting was made on canvas. According to Pacheco's treatise and our results, the canvas was probably first covered with several layers of animal glue, on top of which one or more preparation layers (*preparación* as he calls it; [3,20]) made with Seville's clay ("barro que se usa en Sevilla") and lead white (albayalde) or minium (azarcón) were extended. He also mentions priming (emprimación or emprimadura) referring to a colored priming that could be overlaid on top of the preparation [22]. The existence of the latter cannot be confirmed by XRF alone. This kind of preparation offers already toned painting support that helps in quicker and easier modeling.

The pigment palette known from Pacheco's treatise and other material studies corresponds to our results; nevertheless, the broad use of smalt found in La Inmaculada is not common in his paintings. According to his treatise, he preferred to use "the blue of Santo Domingo" (azul de Santo Domingo), which most studies recognizes as azurite, however it has not been identified conclusively. It could refer to a naturally occurring copper blue from the Spanish Indies, as suggested by Zahira Véliz in her English translation of a part of the Third Book of Pacheco's treatise [22]. He instructs to work on the blue intensity by grounding the pigment more thoroughly for lighter blue tones and leaving larger particles for a darker color [3,20,22]. In some parts he also refers to the "fine ash blues" (azules de cenizas finas), which might be the same pigment, but it is not clear. The blue pigment was generally also chosen to create green hues, which he did not obtain by a simple green pigment, although he suggests including in a painter's palette (verde de montaña y tierras sutiles), which can correspond to malachite and green earth. To obtain a green color, he suggests mixing the blue pigment with a yellow one, often yellow ochre ("buenos ocres") or a yellow lake (amarillo de ancorca, as Pacheco calls the dark yellow tincture of plant origin on a calcium substrate), especially for dark greens. According to his treatise, he also used dark indigo (añil) or purple archil (orchilla) to darken blues [3,20,22]. Archil was obtained from the composite organism lichen, one of the main historic products exported from the

Canary Islands. In this research, we can confirm the use of azurite (or a copper-based blue of Santo Domingo) in the Virgin's blue coat, but cannot identify either the yellow lake or indigo, both being organic. The presence of Cu in the green tones might characterize blue azurite or a copper-based green pigment (perhaps malachite), which could be clarified only by a stratigraphic section.

Pacheco considered flesh tones to be one of the most important parts of a painting and wrote largely about how to achieve a certain tonality: for young women he suggests light flesh tones, obtained by high amount of lead white ("lindo albayalde Venecia"), mixed with some vermilion ("bermellón famoso") and red lake (he mentions carmín), the latter not possible to detect with certainty by XRF in La Inmaculada. Furthermore, he added a small amount of yellow ochre and some azurite, but not a copper-based green pigment, as it was common in Europe for the flesh tones of living people; azurite was used frequently for corpses [3,20,22]. This mixture was confirmed by our study, as seen in Figure 6. Yellow ochre was one of his most used pigments, which he applied generously in flesh tones and in backgrounds; therefore, in this painting we also identified important Fe presence in the background (Figure 8, black line). However, he considered lead–tin yellow (probably identified with genuli in his text; [22]) as the best yellow pigment, creating bright colors; nevertheless, in the present artwork it was rarely found (Figure 7). We also did not find orpiment in this painting, although he mentions it as jalde in his treatise [3,22]. As explained above, organic materials cannot be detected by XRF, but due to high Ca intensity in some pink, red, and dark areas, we proposed a possible use of red lake, yellow lake, and bone black. Red lake was suggested for pink flowers on the lower part of the scene and the Virgin's flesh tones. Pacheco's texts about the achievement of a pink color confirm it, instructing to sketch first with vermilion, overlaying it with lead white and carmine [3,20,22]. As a black pigment, he preferred the use of carbon black; nevertheless, we still suggest the use of bone black in very dark tones, as the shade of the Virgin's hands (Figure 6, blue line). For shades he often used also umber ("sombra de Italia"). His palette corresponds to other great painters of the Golden Area of Spanish painting [18–21], and the pigments found on La Inmaculada uphold it. With our study it was not possible to analyse binders, but from his treatise and previous studies, it is known that he preferred to use linseed oil (aceite de linaza), adding as siccative a vitreous pigment (probably smalt) or litharge (side product of burning lead pigments) and rarely minium (lead red), which could correspond to the presence of Co or Pb in the wide painted surface of the artwork.

Another feature to keep in mind is Pacheco's preparatory drawing. For him, the drawing is the essential part of every painting: "The sure and good grace of the work is in the proper delineation of the entire figure or history because it's certain that all the difficulty of paintings consists in the contours" [22]. However, he believes that drawings of the same size as the final work are not necessary, but suggests composing images per parts, as a mosaic. His bozzettos are first drawn with black stencil, finished with precision using a pen and ink, modeling volumes and shades by different grades of ink dilution. Once the sketch was finished, it was passed on to the canvas, using charcoal or other materials [20]. Therefore, it is surprising that IRR does not reveal much, but when comparing our images with other IRR studies [11–13,20,27], it is clearly quite common. This is explained by Pacheco using dark lines on dark preparation, especially where dark colors were overlaid. With time, on lighter or more transparent areas, some lines can appear, as can be observed on La Inmaculada on the half-moon and the boat (Figure 5a,b). In other IRR studies where the underdrawing was revealed [20,27,28], very sharp black lines with white highlights made with a brush can be observed that perfectly correspond to the colored overlaid paint; this also makes the underdrawing more difficult to discern. His drawings express a very confident, precise, and high-quality artist, who prepared his works very thoroughly, which is why almost no *pentimenti* can be found. Again, La Inmaculada confirms this.

5. Conclusions

The analysis carried out on Pacheco's La Inmaculada panel painting confirms the materials used in the preparation and color layers, as known from the artist's treatise and other studies. They also correspond to the painting procedures of the 17th century "Golden Age" in Spain and especially Seville. Regarding this material characterization, the conservation–restoration process was completed accordingly, and the painting returned to the private area of the Archbishop's Palace in time. Although some questions remain open, there will be no other possibility of performing more analyses on this canvas. Nevertheless, the information gathered with UVF, IRR, and XRF offers an important knowledge that can be used by art historians, restorers, and conservation scientists in the study of Pacheco's works or as a comparison to other Inmaculadas that could be analysed in the future.

Author Contributions: Conceptualization, A.K., J.M.-S., A.G.-O., A.M.-d.-S. and M.Á.R.; methodology, A.K. and J.M.-S.; software, A.K., J.M.-S. and F.J.A.; validation, A.K., J.M.-S. and F.J.A.; formal analysis, A.K., A.G.-O. and A.M.-d.-S.; investigation, A.K., J.M.-S., A.G.-O. and A.M.-d.-S.; resources, A.G.-O. and A.M.-d.-S.; data curation, A.K., J.M.-S. and F.J.A.; writing—original draft preparation, A.K.; writing—review and editing, A.K., J.M.-S., F.J.A., A.G.-O., A.M.-d.-S. and M.Á.R.; visualization, J.M.-S., F.J.A., A.G.-O., A.M.-d.-S. and M.Á.R.; supervision, A.G.-O. and A.M.-d.-S.; project administration, F.J.A. and M.Á.R.; funding acquisition, M.Á.R. All authors have read and agreed to the published version of the manuscript.

Funding: Work supported by the project P18-RT-1877 of Junta de Andalucía (Spain) and by the Margarita Salas contract given by the Spanish Ministry of Universities and funded by the European Union—"NextGenerationEU".

Data Availability Statement: The data presented in this study are available on request from the corresponding author. They are not publicly available due to the intern data policy.

Conflicts of Interest: The authors declare no conflict of interest.

References

1. Valdivieso, E.; Serrera, J.M. *Catálogo de las Pinturas del Palacio Arzobispal de Sevilla;* Editorial Sever Cuesta: Sevilla, Spain, 1979.
2. Valdivieso, E. *Francisco Pacheco;* Caja San Fernando: Sevilla, Spain, 1990.
3. Pacheco, F. *Arte de la Pintura;* Bassegoda i Hugas, B., Ed.; Cátedra: Madrid, Spain, 2001.
4. Escobar Correa, J.G. *Ave Maria Gratia Plena: Oconología e Iconografía de la Inmaculada Concepción;* Universidad Nacional de Colombia: Bogotá, Colombia, 2012.
5. Parejo Delgado, M.J. La iconografía de la Inmaculada Concepción en las parroquias sevillanas. In *La Inmaculada Concepción en España: Religiosidad, Historia y Arte, Symposium Proceedings, Bd. II;* Campos, F.J., Ed.; Real Centro Universitario Escorial-María Cristina, Ediciones Escurialenses San Lorenzo del Escorial: Seville, Spain, 2005.
6. Matteini, M.; Moles, A. *Scienza e Restauro. Metodi D'indagine;* Nardini: Firenze, Italia, 1994.
7. Stuart, B.H. *Analytical Techniques in Materials Conservation;* John Wiley & Sons, Ltd.: Chichester, UK, 2007.
8. Rene de la Rie, E. Fluorescence of Paint and Varnish Layers, Part I–III. *Stud. Conserv.* **1982**, *27*, 1–7, 65–69, 102–108.
9. Hain, M.; Bartl, J.; Jacko, V. Multispectral analysis of cultural heritage artefacts. *Meas. Sci. Rev.* **2003**, *3*, 9–12.
10. Cosentino, A. Notas prácticas sobre a fotografía de ultravioleta para o examen de obras de arte. *Conserv. Patrim.* **2015**, *21*, 53–62. [CrossRef]
11. Van Asperen de Boer, J.R.J. Infrared Reflectography: A Method for the Examination of Paintings. *Appl. Opt.* **1968**, *7*, 1711–1714. [CrossRef] [PubMed]
12. Mairinger, F. The infrared examination of paintings. In *Radiation in Art and Archeometry;* Creagh, B.D.A., Ed.; Elsevier: Amsterdam, The Netherlands, 2000; pp. 40–55.
13. Faries, M. Analytical Capabilities of Infrared Reflectography: An Art Historian's perspective. In *Scientific Examination of Art: Modern Techniques in Conservation and Analysis;* Berrie, B., René de la Rie, E., Hoffman, R., Tomlinson, J., Wiesel, T., Winter, J., Eds.; National Academies Press: Washington, DC, USA, 2005; pp. 87–104.
14. Seccaroni, C.; Moioli, P. *Fluorescenza X: Prontuario per L'analisi XRF Portatile Applicata a Superficie Policrome;* Nardini: Firenze, Italy, 2004.
15. Janssens, K. X-ray based methods of analysis. In *Non-Destructive Microanalysis of Cultural Heritage Materials;* Jannsens, K., Ed.; Elsevier: Amsterdam, The Netherlands, 2004; pp. 129–226.
16. Glinsman, L.D. The practical application of air-path X-ray fluorescence spectrometry in the analysis of museum objects. *Stud. Conserv.* **2005**, *50*, 3–17. [CrossRef]
17. Artioli, G. *Scientific Methods of Cultural Heritage;* University Press: Oxford, UK, 2010.

18. Broquetas, R. *Técnicas y Materiales de la Pintura Española en el Siglo de Oro*; Fundación de Apoyo a la Historia del Arte Hispánico: Madrid, Spain, 2002.
19. Gayo, M.D.; Jover de Celis, M. Evolución de las preparaciones en la pintura de los siglos XVI y XVII en España. *Boletín Mus. Prado* **2010**, *28*, 39–59.
20. Muñoz Rubio, M.V. Francisco Pacheco: Teoría y práctica. In *Pacheco, Teórico, Artista, Maestro*; Muñoz Rubio, M.V., Cano Rivero, I., Eds.; Junta de Andalucía, Consejería de Cultura: Sevilla, Spain, 2016.
21. Križnar, A.; Ager, F.J.; Caliri, C.; Romano, F.P.; Respaldiza, M.R.; Gómez-Morón, M.A.; Núñez, L.; Magdaleno, R. Study of two large-dimension Murillo's paintings by means of macro X-ray fluorescence imaging, point X-ray fluorescence analysis, and stratigraphic studies. *X-ray Spectrom.* **2019**, *48*, 482–489. [CrossRef]
22. Véliz, Z. Francisco Pacheco's Comments on painting in Oil. *Stud. Conserv.* **1982**, *27*, 49–57.
23. Mayer, R. *The Artist's Handbook of Materials and Techniques*, 5th ed.; Viking Penguin: New York, NY, USA, 1991.
24. Wehlte, K. *Werkstoffe und Techniken der Malerei*; Urania: Stuttgart, Germany, 2005.
25. Feller, R.L.; Roy, A.; West Fitzhugh, E.; Berrie, B. (Eds.) *Artists' Pigments: A Handbook of Their History and Characteristics*, 2nd ed.; Archetype Publications: London, UK, 2012.
26. Cocciato, A.; Moens, L.; Vandenabeele, P. On the stability of mediaeval inorganic pigments: A literature review of the effect of climate, material selection, biological activity, analysis and conservation treatments. *Herit. Sci.* **2017**, *5*, 12. [CrossRef]
27. Navarette Prieto, B. Un nuevo dibujo d Francisco Pacheco para su Libro de Retratos: La efigie de Pablo de Céspedes en los Uffizi. *Arch. Español Arte* **2016**, *89*, 77–84. [CrossRef]
28. Navarette Prieto, B. El dibujo como fundamento del pensamiento de Francisco Pacheco: Procesos y principios. In *Pacheco, Teórico, Artista, Maestro*; Muñoz Rubio, M.V., Cano Rivero, I., Eds.; Junta de Andalucía, Consejería de Cultura: Sevilla, Spain, 2016.

minerals

MDPI

Article

Relationship between Rock Porosity and Infrared Cooling Rate in Non-Standard Specimens of Tuffs Used in the Hungarian Cultural Heritage

Simone Mineo [1], Luigi Germinario [2,3], Ákos Török [3] and Giovanna Pappalardo [1,*]

[1] Department of Biological, Geological and Environmental Sciences, University of Catania, Corso Italia 57, 95129 Catania, Italy; simone.mineo@unict.it
[2] Department of Geosciences, University of Padova, Via Gradenigo 6, 35131 Padova, Italy; luigi.germinario@gmail.com
[3] Department of Engineering Geology and Geotechnics, Budapest University of Technology and Economics, Műegyetem rkp. 3, 1111 Budapest, Hungary; torok.akos@emk.bme.hu
* Correspondence: giovanna.pappalardo@unict.it

Abstract: This paper is focused on the application of Infrared Thermography to non-standard rock specimens, in terms of size and deterioration conditions, of Hungarian tuff to monitor their cooling process and to look for a relationship between the rock Cooling Rate Index and the porosity. Literature data agree on the potential of Infrared Thermography for the indirect estimation of rock porosity in fresh specimens through the IRTest, but this technique has never been tested on non-standard specimens. To this purpose, tests on three varieties of Hungarian tuffs were carried out. These materials were selected for their cultural importance linked to their usage as building stones and in other historical applications in Northern Hungary. Tuff specimens underwent a fixed number of salt crystallization cycles. The Cooling Rate Index (CRI) for each specimen was calculated according to the literature experience and correlated to their porosity estimated by water, helium, and mercury intrusion. The results show that the rock cooling process is related to porosity since more porous rocks are characterized by faster cooling. Positive linear trends were achieved for weathered specimens considering 20 min monitoring (CRI_{20}), which is double the time suitable for untreated rocks. The reason should be searched in salt crystallization's effects on the rock texture, paving the way to further studies on this pioneering branch of technological application.

Keywords: Cooling Rate Index; intact rock specimen; porosity; weathering; tuff

1. Introduction

Porosity is a physical property conditioning rock mechanical behavior under stress in terms of strength and deformability [1–5]. The presence of a void network within the rock represents a weakening feature, reducing the contact between grains, lowering rock resistance against axial stresses, and enhancing the rock attitude towards deformation. The solid literature experience in this subject points out that Uniaxial Compressive Strength (UCS) and Young's modulus are inversely proportional to porosity, with statistical trends scattered depending also on the type of voids in the rock, i.e., pores or cracks or both, e.g., [6–8]. Porosity is recognized to be also related to rock durability from the point of view of weathering. In fact, the growth of salt and ice crystals in the pore system and their related stresses can be acknowledged among the major deterioration mechanisms of porous materials [9–15]. This physical property can be referred to as "effective" and "total" depending on whether only the volume of interconnected voids or the total void volume, respectively, is considered. Porosity estimation is therefore relevant for predicting the behavior of rocks used as construction or building materials and specific engineering geological aspects. The International Society for Rock Mechanics (ISRM) and the European Committee for Standardization Comité Européen de Normalisation (CEN) defined the standard procedures to estimate such properties on intact rock specimens following different steps, such as saturation, rock volume estimation, and the pycnometer analysis on a pulverized specimen fraction [16,17]. With respect to the effective porosity, the results achieved by such procedures refer to the pore volume that can be filled with water, and this kind of porosity will be defined from now on in the text, as n'_{H_2O}. Other approaches are based on the use of gases (helium, nitrogen, carbon dioxide, etc.) or liquids (mercury), which are forced to fill the rock voids under a pressure gradient for the porosity estimation (e.g., n_{He} and n'_{Hg}). In any case, these operations require rock specimen destruction to prepare either a powder or a small rock fragment. Still, other techniques are based on microscopic image analysis, but their measurement range and accuracy are relatively limited and may still require sample preparation for best results [18]. In the last decade, an innovative procedure for the non-destructive estimation of rock porosity through Infrared Thermography (IRT) was presented [19–22]. The IRTest is based on the positive correlation existing between the cooling rate of an oven-heated rock specimen and its porosity, representing a promising quick experimental procedure. In the literature, IRTests have been performed on regularly shaped intact rock specimens, sized according to international standards, to estimate both total porosity and n'_{H_2O}, proving that the cooling rate within the first 10 min of monitoring (CRI_{10}) is a promising index for the indirect prediction of such physical property. Nevertheless, the IRTest has never been tested on non-standard specimens in terms of size and deterioration conditions. In this perspective, the aim of this paper is to find out whether this can provide information on the rock porosity of small rock specimens, even after artificial aging by salt crystallization cycles. These experimental requirements may reflect, among other things, those for testing historical materials in cultural heritage, which often entail restricted sampling and a bad state of conservation. To this purpose, tests on three varieties of Hungarian tuffs were carried out. These materials were selected for their cultural importance linked to their usage as medieval building stones and in other historical applications, especially in the region of Northern Hungary. Tuff specimens were grouped into six groups, each undergoing a fixed number of salt crystallization cycles. For each specimen, the Cooling Rate Index (CRI) was calculated according to the literature experience, and then porosity was estimated under water, helium, and/or mercury intrusion. The data collected were finally used for statistical correlations to shed light on the suitability of the IRTest for the porosity assessment of non-standard rock specimens.

2. Materials

The materials investigated herein are pyroclastic rocks historically used as building and ornamental stones in Hungary, named after the towns of Bogács, Demjén, and Sirok in the Eger region (Northern Hungary), where they are quarried. These tuffs have a centuries-old tradition of usage that commenced as early as the Middle Ages. Remarkable examples of historical monuments made of tuff include Christian fortifications and churches, for instance, the 13th-century castle and the 19th-century Baroque cathedral of Eger, as well as Ottoman mosques and baths (17th century) (Figure 1). These stones have also had more modest, although widespread, applications in vernacular architecture, for building houses, cave dwellings, wine cellars, burials, etc. [23,24].

Considering their geological background, they belong to a sequence of Miocene welded and unwelded pumiceous tuffs with a prevailing high-K rhyolitic composition, formed between 18.2 and 14.4 Ma during the tectonic evolution of the Carpathian-Pannonian Region [25]. They are classifiable as pumice tuffs with a pseudo-porphyritic texture, glassy groundmass, and a mineralogical assemblage composed mainly of plagioclase, biotite, quartz, and occasionally, amphibole and pyroxene; swelling clay minerals are the most significant secondary phases.

Demjén tuff is the most porous and soft, with compressive and tensile strengths of 10 and 1 MPa, respectively. Bogács and Sirok tuffs have better physical and mechanical properties, with compressive strength and tensile strength in the ranges of 26–28 MPa and

3–4 MPa. For all three tuff varieties, water saturation may cause an extreme deterioration of the mechanical performance [23].

Figure 1. Examples of the historical usage of tuffs in the town of Eger: the Medieval castle and the Ottoman minaret. The insets show the location of the study quarry sites (a few kilometers apart) and the macroscopic appearance of Demjén tuff used in the monuments.

3. Methods

The methodological approach set for this study can be summarized by 4 main points: (1) salt crystallization tests; (2) IRTest for the calculation of the Cooling Rate Indexes; (3) rock porosity estimation under water, helium, and mercury intrusion; (4) statistical data processing.

3.1. Salt Crystallization Tests

Salt crystallization tests were performed to reproduce the rock degradation effects naturally arising from the weathering phenomena involving in-pore water circulation. Tuff specimens had a non-standard shape, being cylinders with a 50 mm diameter and a 30 mm height. A total of 72 specimens were prepared. These were then grouped into 6 sets, each undergoing a different number of crystallization cycles (0, 2, 4, 6, 10, 23), with 1 set serving as an untreated reference (Cycle 0). The different steps were chosen based on how quickly the observed state of conservation changed, so as to isolate representative samples for different decay stages for the following IRTests. According to the standard test method EN 12370 (1999), each salt crystallization cycle consisted of the following phases: 2 h total immersion in a 14% Na_2SO_4 aqueous solution, minimum 16 h drying at 105 °C, and 2 h cooling. After each cycle, the sample mass and macroscopic changes were recorded.

3.2. Infrared Cooling Monitoring and CRI Calculation

The IRT technique relies on the principle that every material at a temperature above absolute zero emits thermal radiation. This is characterized by a wavelength mainly falling within the infrared band of the electromagnetic spectrum [26], and it is proportional to the temperature of the emitting body, according to the Stefan–Boltzmann law. Therefore, the surface temperature of a body can be easily estimated by a thermal camera, which is a device operating within the range of wavelengths as long as 13–14 μm. The use of IRT for rock characterization has increasingly become the subject of numerous scientific studies focused either on rock deterioration and fracturing state [11,27–30] or on rocky outcrops [31–36].

Recent literature studies shed light on the positive correlation between rock cooling, monitored by IRT, and porosity [19–22]. Starting from the hypothesis that the presence of voids within the rock texture could affect its thermal behavior, a series of tests on numerous intact rock specimens were performed. These proved that the greater the porosity of the rock, the faster its cooling is.

The IRTest on intact rock specimens is carried out according to the following four main stages:

(1) Oven-heating of the rock specimens for 24 h at 70 ± 5 °C.

(2) Monitoring of the rock cooling by acquiring regular IRT images of the upper face of the specimens, which are placed on a smooth laminate worktop in a temperature-controlled room (25 ± 1 °C), for a defined time interval. The initial IRT frame is acquired when the cooling begins, thus representing the reference for the following stage.

(3) Post-processing of the IRT images in order to define the average surface temperature of the framed specimen face; this is the average temperature value calculated within a specific area of the IR image, set to coincide with the specimen's upper face.

(4) According to Equation (1), the Cooling Rate Index (CRI) of the rock is calculated for a defined monitoring time window. CRI is expressed by the ratio between the difference of temperature between two measurements (ΔT) within a specific time window (Δt).

$$\mathrm{CRI} = \Delta\mathrm{T}/\Delta\mathrm{t} \quad (1)$$

Literature outcomes agree that the best relationship between rock porosity (estimated by buoyancy in water) and rock cooling is achieved by calculating the CRI within a 10 min time window starting from when the cooling begins (CRI_{10}). A satisfactory correlation was found for rocks belonging to different geological formations and for different shapes and sizes of specimens [19–22].

In this paper, the IRTest was experimentally carried out, on the same specimens prepared for salt crystallization (see Section 3.1), by using a high-sensitivity infrared thermal imaging camera with a 320 × 240 pixels infrared resolution and operating in a range of temperatures from −20 to 650 °C (with ±2 °C accuracy) and a thermal sensitivity < 0.04 °C. The rock emissivity was set to 0.975, suitable for the rock type tested according to the literature [37].

Rock IR monitoring had a 5 min cadence, thus allowing the final computation of CRI within the first 5, 10, 15, and 20 min from the cooling beginning.

Thermal images were post-processed through software capable of handling radiometric images and extracting the average surface temperature values corresponding to the framed rock specimen face (Figure 2).

3.3. Rock Porosity Estimation

The porosity of the reference group of tuff specimens (Cycle 0) was estimated according to EN1936:2006 by the long-established Archimedes buoyancy method on 16 specimens. This method relies on the measure of three parameters of a single specimen: the weight of the dry specimen (w_d); the weight in air of the specimen fully saturated with water (w_s); and the weight of the saturated specimen suspended and fully immersed in water (the Archimedes weight w_h) [38]. From these weights, and after determining the real rock density (ρ_r) through the pycnometer method, total (n) and effective (n'_{H_2O}) porosities can be calculated according to Equations (2) and (3), where ρ_b is the bulk density.

$$n = \left(1 - \frac{\rho_b}{\rho_r}\right)100 \quad (2)$$

$$n'_{H_2O} = \frac{w_s - w_d}{w_s - w_h} \quad (3)$$

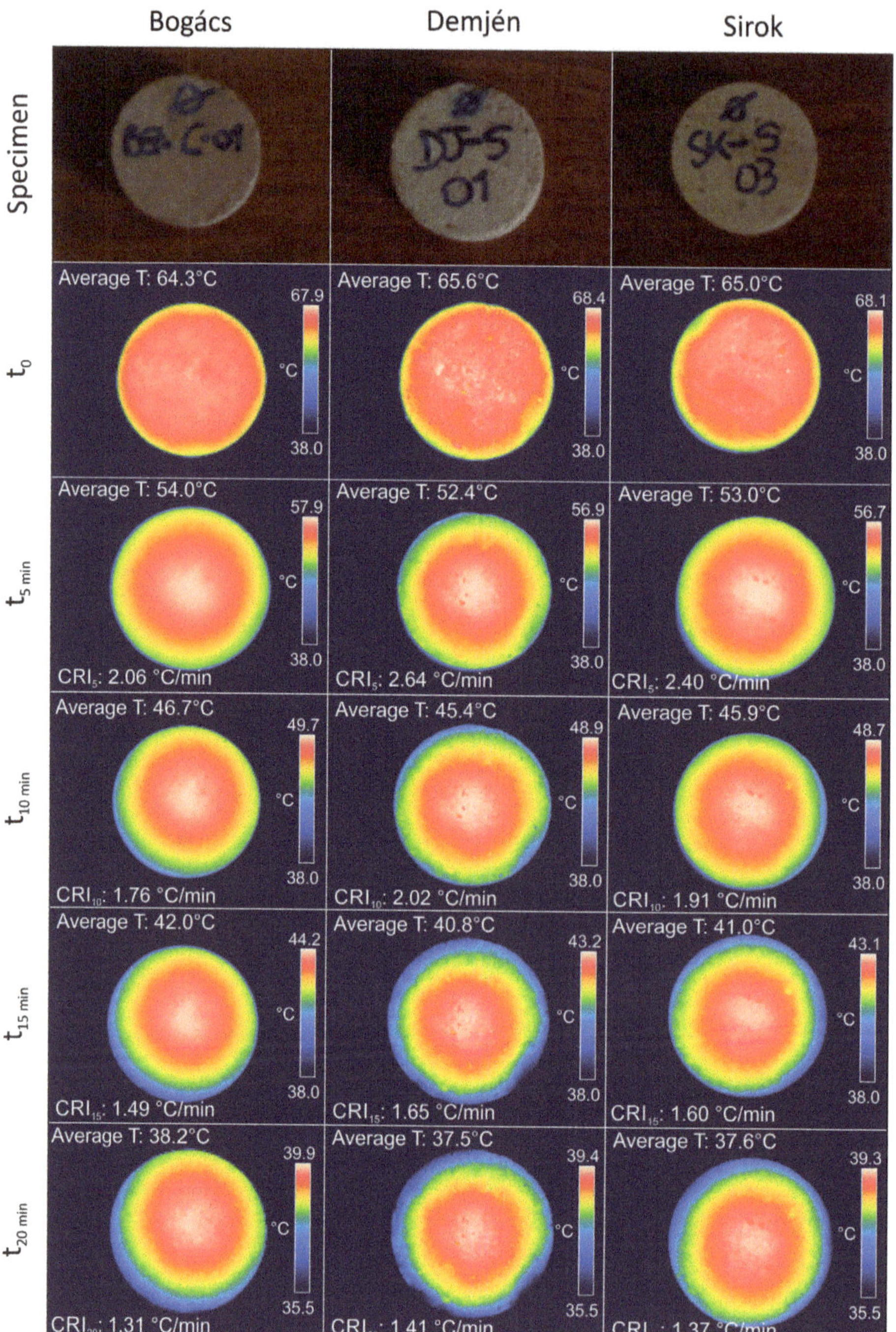

Figure 2. Summary of the IR monitoring for selected representative specimens of each untreated tuff variety.

Some specimens started showing a loose structure from the second cycle, sometimes crumbly even to the touch for the most advanced cycles. This made them unsuitable for saturation with water. Therefore, to avoid errors in the porosity estimation due to both possible salt crystal dissolution and partial loss of specimen mass, porosity was computed by a multianalytical approach using structured-light 3D scanning, helium pycnometry, and mercury intrusion porosimetry (MIP) on 10 representative weathered samples from Cycles 2, 4, 6, 10, and 23.

By structured-light 3D scanning, a Scan-in-a-box FX scanner was combined with the software IDEA 1.1 SR 8 FX for building 3D digital models of the samples and calculating their volume, averaging from three independent measurements. The weighing of the samples provided their bulk density.

By helium pycnometry, a Quantachrome Pentapyc 5200e PPY-30T pycnometer was operated at 25 °C constant temperature and 131 kPa maximum pressure for determining the matrix volume and, indirectly, the matrix density of the samples, averaged among three measurements. Total porosity and pore volume were also calculated.

By MIP, a Quantachrome Poremaster 60 GT porosimeter was used at low pressure (1.5–350 kPa, for pore-throat diameters of 382–4.26 μm) and high pressure (140 kPa–420 MPa, for pore-throat diameters of 10.66–0.004 μm), with two repeated measurements per pressure stage, for determining open porosity, pore-size distribution, and specific surface area.

3.4. Statistical Data Processing

CRIs within 5, 10, 15, and 20 min of cooling were plotted against the n and n'_{H_2O} values for the untreated specimens, measured according to the EN1936:2006 specifications. A simple regression analysis was carried out, and the best-fit curves and the coefficients of determination (R^2) were calculated by the "least squares" method to determine if the correlation between porosity and CRI could be ensured. The achieved statistical trends were then compared to the literature trends [21], referring to specimens similar in shape and size to those tested herein, to prove the reliability of the testing principle. For the other tuff groups (from Cycle 2), CRIs were correlated to n_{He} and n'_{Hg} porosity to find out if significant correlations can be achieved as well.

4. Results of Salt Crystallization Tests

A summary of the results of the salt crystallization tests is shown in Figure 3.

Demjén tuff has the lowest durability to salt crystallization, crumbling already during the first aging cycles. Differential erosion is the main weathering pattern, highlighting the faster disintegration of the groundmass, compared to the slower detachment of the coarser-size phenocrysts.

The resistance of Bogács and Sirok tuffs is distinctively higher instead. Most specimens show negligible macroscopic changes and mass variations (excluding the weight increases due to the continuous salt loading) throughout the experimental program. Bogács tuffs record just the formation of a few voids on the surface and a slight, occasional peeling, which becomes more evident during the very last cycles. On the other hand, several specimens of Sirok tuff start deteriorating after the 5th cycle, with traces of differential erosion, disintegration, and microcracking, leading to mass decreases up to ~10%. These become even clearer during the following cycles, reaching drops of ~25%–30% for some specimens at the end of the tests. The strength variability observed in Sirok tuffs, dependent on changing textural characteristics (e.g., different amounts of pumice clasts, glassy groundmass, coarse phenocrysts and lithoclasts, welding degree), is in agreement with previous experimental observations [23].

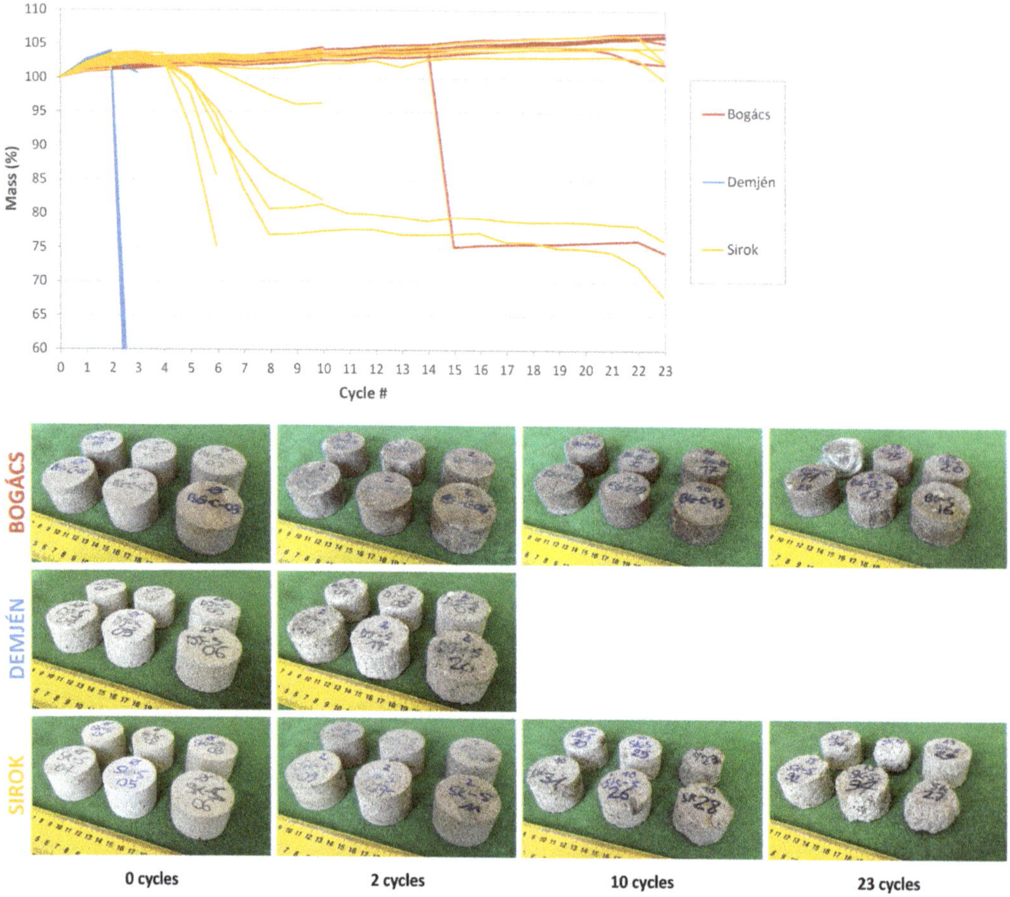

Figure 3. Mass changes of the three tuff varieties during the salt crystallization tests, with photographs of the specimens after different cycles, showing the weathering evolution (the curves refer to all the tested specimens).

5. Results of Porosity Estimation

The initial rock physical characterization, carried out through the method of Archimedes buoyancy in water, returned dissimilar porosity characteristics among the tested tuff varieties. In particular, Demjén tuff has the highest total porosity and the lowest bulk density (Table 1), statistically affected by the low standard deviation. This is an index of textural uniformity. On the other hand, Bogács tuff is characterized by the lowest total porosity and the highest bulk density. Among the studied tuffs, Sirok shows the highest standard deviation for total and effective porosity (Table 1). The degree of pore interconnection is the highest for Demjén, with an effective porosity of 72.6% of the rock volume, and the lowest for Bogács, with an effective porosity representing only 54% of the total porosity. The high pore volume and interconnection in Demjén tuff is associated with a distinctively higher water absorption and salt penetration, which explains the much lower durability to salt crystallization, especially since pore-size distribution is quite similar to Bogács tuff. In addition, the lower tensile strength implies a poorer mechanical performance when in-pore stresses are generated during salt crystallization and dissolution [23].

Table 1. Average (±standard deviation) values of bulk density, total, and effective porosity estimated on untreated tuff varieties.

Tuff Variety	Total Porosity (%)	Effective Porosity (%)	Bulk Density (g/cm^3)
Bogács	29.37 (±2.1)	15.86 (±0.7)	1.78 (±0.05)
Demjén	37.15 (±0.5)	26.97 (±0.4)	1.51 (±0.01)
Sirok	32.93 (±3.7)	22.15 (±2.2)	1.64 (±0.09)

As the salt crystallization tests go on, considerations can be made only on the strongest specimens of Bogács and Sirok, although no relevant differences between the two tuffs can be highlighted as the cycles progress, with the relevant n'_{Hg} and n_{He} porosity values keeping almost constant trends. This can be explained by the progressive salt accumulation within the rock pores, which balances the ongoing rock weathering, hiding its effects on the porosity. This consideration is strengthened by the positive trend of mass change shown in Figure 3.

Moreover, considerations about the possible changes in pore-size distribution as salt weathering progresses can be presented only about Bogács and Sirok tuffs, supported by the full MIP datasets in Supplementary File S1. Bogács tuff shows no significant changes in pore-size distribution, characterized by capillary pores in the approximate range between 1 and 10 μm, which stays virtually the same during the continuation of the tests. Sirok tuff has a higher concentration of micropores, with broader and multimodal pore-size distribution curves, with peaks mainly in the approximate range between 0.01 and 1 μm; no clear trend of changing pore-size distribution during the experimental tests is noticeable, apart from a certain degree of disturbance (i.e., higher pore-size variability) from the 4th salt crystallization cycle onwards in Table 2. However, since the number of capillary pores (those most involved in the mechanisms of water absorption and movement) is lower in Sirok tuff, the prosecution of the salt tests with a higher number of cycles might eventually highlight the higher durability of this material [23].

Table 2. Results of the porosity measurements of the tuff specimens subjected to different numbers of salt crystallization cycles. The full pore-size distributions by MIP are included in Supplementary File S1.

Tuff Variety	Sample ID	Cycle Number	He Pycnometry				MIP		
			Matrix Density (g/cm^3)	Bulk Density (g/cm^3)	Total Porosity (%)	Specific Pore Volume (cm^3/g)	Main Peaks of Pore-Size Distribution Curves (μm)	Open Porosity (%)	Specific Surface Area (m^2/g)
Bogács	BG_C_8	4	2.5713 (±0.0003)	1.818 (±0.001)	29.29 (±0.05)	0.161 (±3.5 × 10^{-4})	4.99 | 2.88	25.67	10.54
	BG_C_S_19	6	2.5207 (±0.0004)	1.809 (±0.001)	28.22 (±0.02)	0.156 (±1.6 × 10^{-4})	4.73 | 3.11	23.80	11.15
	BG_C_09	10	2.5414 (±0.0001)	1.835 (±0.001)	27.79 (±0.04)	0.151 (±3.2 × 10^{-4})	4.74 | 2.96 | 0.07	26.29	9.32
	BG_C_11	23	2.5343 (±0.0003)	1.897 (±0.001)	25.13 (±0.04)	0.132 (±2.9 × 10^{-4})	5.53 | 4.99	19.81	9.42
Demjén	DJ_S_09	2	2.4977 (±0.0004)	1.474 (±0.002)	40.97 (±0.06)	0.278 (±7.2 × 10^{-4})	5.99	31.66	3.38
Sirok	SK_S_07	2	2.4983 (±0.0010)	1.685 (±0.002)	32.57 (±0.10)	0.193 (±8.5 × 10^{-4})	12.2 | 0.10	30.42	13.50
	SK_35	4	2.4617 (±0.0004)	1.709 (±0.002)	30.58 (±0.10)	0.179 (±8.2 × 10^{-4})	1.04 | 0.08 | 0.02	22.76	15.20
	SK_S_21	6	2.4788 (±0.0005)	1.713 (±0.001)	30.87 (±0.04)	0.180 (±2.8 × 10^{-4})	0.14	26.39	10.76
	SK_S_31	10	2.4786 (±0.0009)	1.739 (±0.001)	29.83 (±0.04)	0.172 (±2.5 × 10^{-4})	1.04 | 0.02	23.60	15.82
	SK_S_34	23	2.4546 (±0.0009)	1.763 (±0.002)	28.16 (±0.07)	0.160 (±4.8 × 10^{-4})	0.13	23.41	12.34

6. Results of IR Cooling Monitoring

The analysis of the thermal images acquired for the untreated tested specimens (Cycle 0) during the cooling monitoring (Figure 2) shows that the rock cooling proceeds towards the inner rock in a generally uniform way, according to literature accounts [21], thanks to the circular geometry of the specimen' face. Thermal images keep a good definition for the entire monitoring time windows, allowing the good identification of the rock face and sampling its average surface temperature (Figure 2).

According to the estimated temperature values and the calculated cooling rates, the untreated specimens of the Demjén variety were characterized by the highest CRIs, suggesting the fastest cooling among the tested groups, followed by Sirok and Bogács (Figure 4a). For all the cases, the standard deviation is low, thus suggesting the reliability of the considered average values. The highest CRIs are referred to as the 5 min time window, when the temperature difference between the specimen and the laboratory environment is maximum. This is in accordance with Newton's law of cooling and previous literature accounts on this test [19]. Therefore, the cooling speed is maximum at the initial stages of the test and tends to slow down as time passes, as shown by the progressively decreasing angle of the cooling curves (Figure 4b). By analyzing the representative cooling curves reported in Figure 4b, it can be stated that the Demjén variety is characterized by the fastest cooling, considering the common initial stage and the lowest surface temperature reached at the end of the monitoring. This trend is more apparent even focusing on the initial segment, used to calculate CRI_5, which is steeper for Demjén, thus suggesting the highest cooling rate within the first 5 min of monitoring. Similarly, the less steep curve is related to the Bogács variety, thus suggesting a lower cooling rate, and supporting the average data reported in Figure 4a.

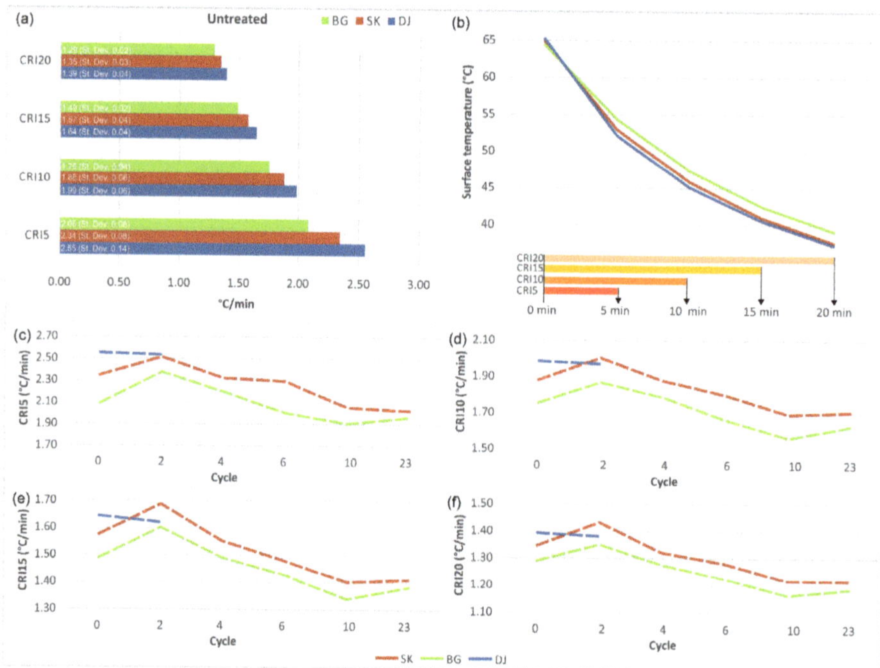

Figure 4. CRI values and cooling curves of tuffs: (**a**) comparison between the average CRI values measured for the three tuff varieties in untreated conditions; (**b**) cooling curves of selected representative untreated specimens belonging to each group; average trends of (**c**) CRI_5, (**d**) CRI_{10}, (**e**) CRI_{15}, and (**f**) CRI_{20} trend with cycle progression. Key: SK—Sirok, BG—Bogács, DJ—Demjén.

By looking at the average CRI trend as the cycles progress (the standard deviation related to each average value is generally <0.1), an initial increase can be reported, with respect to Sirok and Bogács, between Cycle 0 and Cycle 2 (Figure 4c–f). This is about 6%–7% and 5%–12% for Sirok and Bogács, respectively. On the other hand, on average, a CRI decrease of around 1% was found for Demjén (Figure 4c–f). It has to be noted that after the 2nd cycle, no Demjén specimen resisted the crystallization tests, thus the following analysis is referred only to Bogács and Sirok. The related mean CRIs tend to decrease according to an almost constant trend until the 10th cycle. Only the average CRI_5 of Sirok measured between the 4th and the 6th shows a minimal difference (Figure 4c). From the 10th to the 23rd cycle, Sirok's CRIs keep an almost constant value, while Bogács's values mark a slight increase of about 2%–4%.

7. Statistical Correlations

7.1. CRI and Rock Porosity on Untreated (Cycle 0) Specimens

To shed light on the statistical correlation between CRI and rock porosity, experimental cooling data referring to Cycle 0 were plotted against both n and n'_{H_2O} values (Figure 5). Scatter plots show a positive, linear trend between the variables, proving that the fastest rock cooling is related to the most porous rocks. This outcome, verified herein for the tested tuffs' total and effective porosity, is in accordance with literature accounts [19–22].

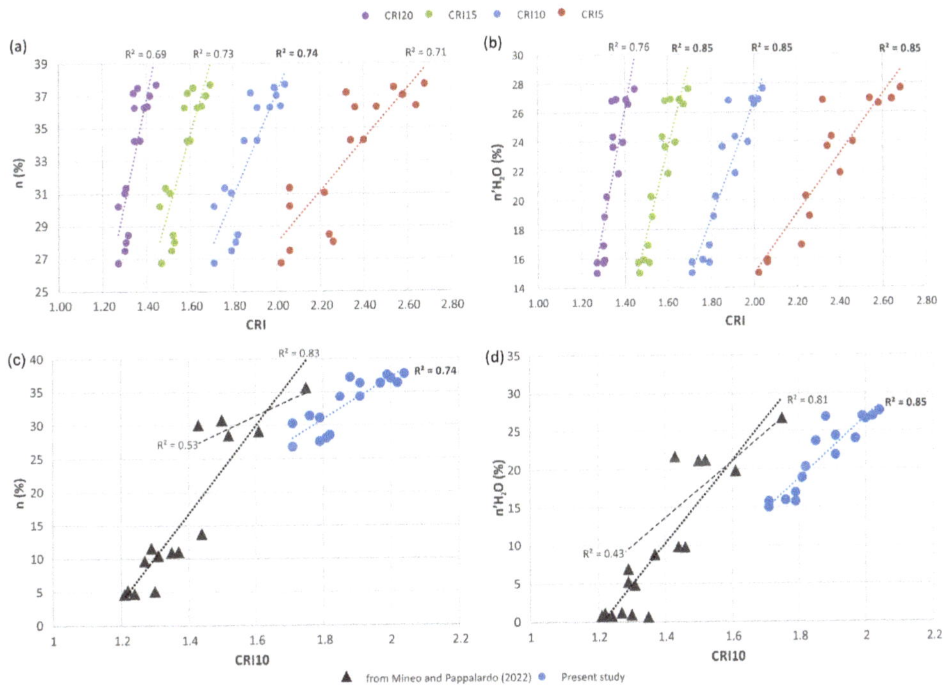

Figure 5. Statistical relationship between thermal cooling and rock porosity for untreated (Cycle 0) specimens: (**a**) correlation diagram CRI vs. total porosity; (**b**) correlation diagram CRI vs. effective porosity; (**c**) CRI_{10} vs. total porosity: comparison between data from this study and literature [21] (dotted lines are the best fit of the whole sets, while the dashed black line is the best fit of literature data involving only specimens with $n > 25\%$ for a more specific comparison); (**d**) CRI_{10} vs. effective porosity: comparison between data from this study and literature [21] (dotted lines are the best fit of the whole sets, while the dashed black line is the best fit of literature data involving only specimens with $n'_{H_2O} \geq 10\%$ for a more specific comparison).

The best correlation of total porosities is achieved by CRI_{10}, with a trend characterized by a 0.74 R^2 against the minimum value of 0.69 belonging to the correlation with CRI_{20} (Figure 5a). On the other hand, the best statistical correlations involving n'_{H_2O} refer to CRI_5, CRI_{10}, and CRI_{15} being characterized by a 0.85 R^2 (Figure 5b). Based on these outcomes, the potential of CRI_{10} for rock porosity prediction is herein strengthened according to Equations (4) and (5). These describe the best trends achieved with the non-standard cylindrical specimens tested for this study, which had a 50 mm diameter and a 30 mm height (indicative volume of 59 cm^3). This is significant, since there is no literature experience on such specimen size. The only available literature data of rock specimens with a somewhat similar size, tested by IRTest, are related to 1:1 small rock cylinders with a 54 mm diameter (indicative volume of 124 cm^3) [21].

It is, therefore, valuable to compare such data to evaluate their scientific soundness. To this purpose, the experimental CRI_{10} statistical plots of the Hungarian tuffs were overlapped with those published in the literature [21] related to the 1:1 small cylinders, bearing in mind the volume difference (Figure 5c,d). It can be noted that the rocks tested herein are, indeed, characterized by a higher cooling rate due to their smaller volume. Nevertheless, a balanced trend comparison should be carried out by taking into account only the specimens showing similar porosity values. Therefore, by focusing the analysis only on the greatest porosity ranges ($n > 25\%$ and $n'_{H_2O} \geq 10\%$), the statistical trends appear sub-parallel, and the highest R^2 values belong to the Hungarian tuffs tested for this study in both n and n'_{H_2O} cases (Figure 5c,d).

$$n\ (\%) = 31.58\ CRI_{10} - 26 \quad (4)$$

$$n'_{H_2O}(\%) = 40.13\ CRI_{10} - 53.45 \quad (5)$$

7.2. CRI and Rock Porosity after Salt Crystallization Cycles

Similarly to what was carried out for the fresh samples, a simple regression analysis between CRI and rock porosity was performed to determine if the statistical relationship persists for weathered rocks (Figure 6a,b). In this case, the CRIs calculated for the rocks belonging to Cycle 2 were taken into account. The results show that the statistical trends described by CRI and n fail to equal the previous ones in terms of statistical goodness (Figure 6a). In fact, more scattered plots were achieved, with maximum R^2 values of 0.53. The worst correlation (R^2 0.23) was found for CRI_5, while the best one refers to CRI_{10}. A similar outcome was outlined for n'_{H_2O}, although with slightly higher R^2 values, with particular reference to CRI_{15} (Figure 6b).

Considering the satisfactory correlations related to the untreated specimens, along with the available literature experience, the poor statistical significance achieved in this case casts doubt on the reliability of the procedure used to estimate rock porosity for the weathered specimens. In fact, the buoyancy method requires the specimen immersion in water for at least 24 h, to achieve saturation. This procedure likely affected the measurements in two ways: (1) weathered rocks tend to gain a looser structure, favoring a certain loss of mass during immersion; (2) salt crystals precipitated within the rock pores could likely be involved in dissolution processes when the rock is saturated. Both circumstances lead to possible dissimilarities between wet and dry specimen masses for the porosity calculation, likely making the so-measured porosity values unrepresentative. Therefore, with the aim of looking for a better statistical relation for such non-standard rock specimens, the n'_{Hg} and n_{He} porosity values, estimated on fragments belonging to 10 selected rock specimens (see Section 3.3), were taken into account. These data were plotted against the CRI values, regardless of the salt crystallization cycle (from Cycle 2 to Cycle 23), and the best fit was achieved for the correlations involving CRI_{20} (Figure 6c,d; Equations (6) and (7)).

$$n'_{Hg} = 30\ CRI_{20} - 12.6 \quad (6)$$

$$n_{He} = 36.6\, CRI_{20} - 15.95 \tag{7}$$

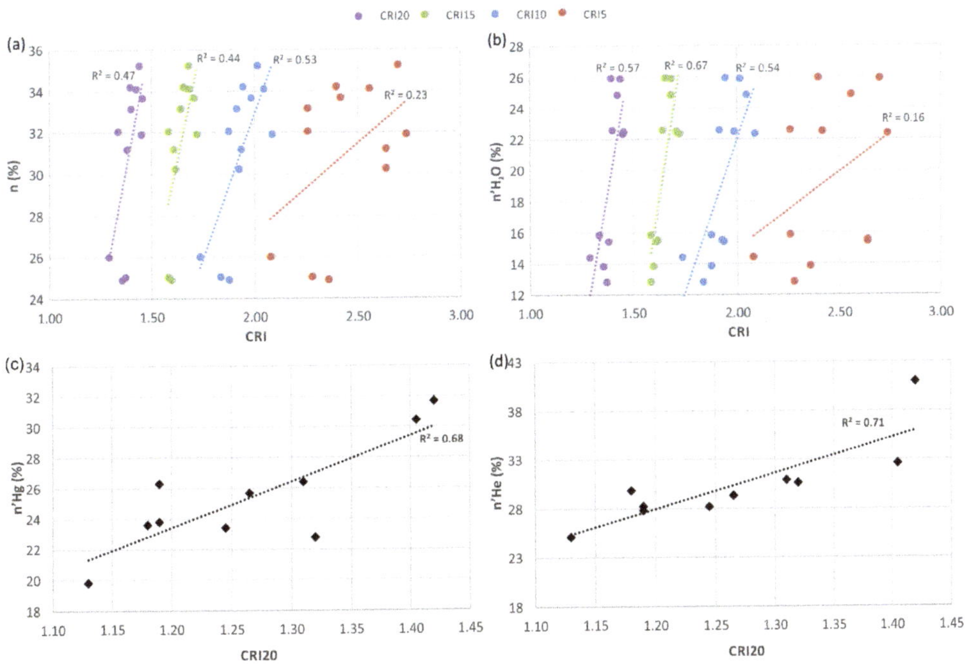

Figure 6. Statistical relationship between thermal cooling and rock total porosity, calculated through the buoyancy method, for the specimens subjected to 2 cycles (**a**); statistical relationship between thermal cooling and rock n'_{H_2O} for the specimens subjected to 2 cycles (**b**); best statistical trend achieved by plotting n'_{Hg} vs. CRI_{20} (**c**); best statistical trend achieved by plotting n_{He} vs. CRI_{20} (**d**).

8. Discussion and Conclusions

The results achieved by this study confirm that the rock cooling process, monitored by IRT through the IRTest, is related to porosity. This is demonstrated by satisfactory linear statistical trends obtained by plotting the rock porosity against the calculated CRIs. It was confirmed that the cooling of rock is directly proportional to the rock volume occupied by voids, concerning both total and effective porosity. In fact, the higher the rock porosity, the faster its cooling. According to the literature background on rock properties and Newton's law of cooling, the cooling velocity of oven-heated rocks is characterized by the highest rates during the first minutes. Then, it tends to slow down as the temperature difference between the rock and the surrounding environment gets smaller. This is confirmed by the trend shown by the cooling curves taken as examples in Figure 4b. The results show also that the cooling rate measured within the first 10 min of cooling (CRI_{10}), which is known in the literature to be a good porosity predictor [19–22], keeps its prediction potential thanks to the best correlations achieved with n and n'_{H_2O}. Its role was strengthened herein by comparing literature data and the experimental results referring to the untreated Hungarian tuffs specimens (Figure 5c,d).

A further consideration is for the specimens' size; in fact, rocks tested for this study had about half of the volume of the specimens from the study taken as Reference [21]. It resulted that the rocks tested herein are characterized by a higher cooling rate due to their smaller volume, underlying that the rock volume plays a key role in the cooling speed, with the highest rates related to lower volumes. Although this study demonstrated that the CRI works fine also for such small rock specimens, there is the need of referring to standardized sample dimensions in the perspective of the potential standardization of such approaches.

Besides the considerations on the untreated specimens, which are essential to have the IRTest validated even for non-standard-sized rock specimens, interesting discussion points arise from the tests carried out on rocks treated by salt crystallization. From the 2nd to the 23rd cycle, n'_{Hg} and n_{He} were taken as a reference for the pioneering correlation against CRI. Positive linear trends were achieved by considering the 20 min time window (CRI_{20}) for all the selected specimens (Figure 6c,d). This is double the most suitable time window for untreated rocks (10 min) and the reason for that should be searched in the effects that salt crystallization has within the rock texture. This aspect needs to be further investigated, even questioning if an application for the specific IRTest can be posted in terms of crystallization cycles. Figure 4c,f show that CRIs between the 10th and the 23rd cycle remain almost constant for Sirok and are affected by an inverse trend for Bogács. However, the good match between rock cooling and Hg-He porosity suggests that porosity plays a leading role in rock cooling and that IR monitoring is a useful approach for its indirect estimation, even for weathered rocks.

Although needing further experimentation, the application of this method to stone materials used in the Hungarian cultural heritage opens interesting possibilities for the imaging and textural characterization of historical materials. At this stage of development, the technique provides a promising perspective in the frame of stone testing for cultural heritage, although available results refer only to the laboratory setting. Moreover, it is totally non-destructive, since it works with temperatures that do not cause significant changes in structural water content, irreversible strains, or mineralogical transformations. In the frame of technological development, more research is needed to find and set a comparable procedure that could be extended for in situ tests.

Supplementary Materials: The following supporting information can be downloaded at: https://www.mdpi.com/article/10.3390/min13081100/s1, Supplementary File S1: MIP pore size distribution.

Author Contributions: Conceptualization, infrared thermography, physical rock characterization S.M. and G.P.; rock sampling, salt crystallization tests and porosimetric tests L.G. and Á.T.; writing—original draft preparation, S.M. and L.G. writing—review and editing, funding acquisition G.P. and Á.T. All authors have read and agreed to the published version of the manuscript.

Funding: This research was financially supported by Università di Catania PIACERI Linea 2 "CH2V—Cultural Heritage Hazard and Vulnerability" project, scientific responsible Giovanna Pappalardo. The support provided by the Ministry of Culture and Innovation of Hungary from the National Research, Development and Innovation Fund, financed under the TKP2021-NVA funding scheme (Project No. TKP-6-6/PALY-2021) to Ákos Török, is acknowledged.

Acknowledgments: This work is published in the frame of the Agreement of Scientific Cooperation between the University of Catania and the Budapest University of Technology and Economy coordinated by Giovanna Pappalardo and Ákos Török. The rock physical characterization and IRTests were performed at the *Laboratorio di Geologia Applicata* of the University of Catania, Department of Biological, Geological and Environmental Sciences. The thermal camera was provided by the *Laboratorio di Analisi non Distruttive* of the same University. Salt crystallization tests, structured-light 3D scanning, helium pycnometry, and mercury intrusion porosimetry were performed at the Budapest University of Technology and Economics.

Conflicts of Interest: The authors declare no conflict of interest.

References

1. Hatzor, Y.H.; Palchik, V. A Microstructure-Based Failure Criterion for Aminadav Dolomites. *Int. J. Rock Mech. Min. Sci.* **1998**, *35*, 797–805. [CrossRef]
2. Vernik, L.; Bruno, M.; Bovberg, C. Empirical Relations between Compressive Strength and Porosity of Siliciclastic Rocks. *Int. J. Rock Mech. Min. Sci. Geomech. Abstr.* **1993**, *30*, 677–680. [CrossRef]

3. Al-Harthi, A.A.; Al-Amri, R.M.; Shehata, W.M. The Porosity and Engineering Properties of Vesicular Basalt in Saudi Arabia. *Eng. Geol.* **1999**, *54*, 313–320. [CrossRef]
4. Chang, C.; Zoback, M.D.; Khaksar, A. Empirical Relations between Rock Strength and Physical Properties in Sedimentary Rocks. *J. Pet. Sci. Eng.* **2006**, *51*, 223–237. [CrossRef]
5. Basu, A.; Celestino, T.B.; Bortolucci, A.A. Evaluation of Rock Mechanical Behaviors under Uniaxial Compression with Reference to Assessed Weathering Grades. *Rock Mech Rock Eng* **2009**, *42*, 73–93. [CrossRef]
6. Palchik, V. Influence of Porosity and Elastic Modulus on Uniaxial Compressive Strength in Soft Brittle Porous Sandstones. *Rock Mech. Rock Eng.* **1999**, *32*, 303–309. [CrossRef]
7. Pappalardo, G.; Mineo, S. Static Elastic Modulus of Rocks Predicted through Regression Models and Artificial Neural Network. *Eng. Geol.* **2022**, *308*, 106829. [CrossRef]
8. Sousa, L.M.O.; Suárez Del Río, L.M.; Calleja, L.; Ruiz De Argandoña, V.G.; Rey, A.R. Influence of Microfractures and Porosity on the Physico-Mechanical Properties and Weathering of Ornamental Granites. *Eng. Geol.* **2005**, *77*, 153–168. [CrossRef]
9. Thaulow, N.; Sahu, S. Mechanism of Concrete Deterioration Due to Salt Crystallization. *Mater. Charact.* **2004**, *53*, 123–127. [CrossRef]
10. Benavente, D.; Sanchez-Moral, S.; Fernandez-Cortes, A.; Cañaveras, J.C.; Elez, J.; Saiz-Jimenez, C. Salt Damage and Microclimate in the Postumius Tomb, Roman Necropolis of Carmona, Spain. *Environ. Earth Sci.* **2011**, *63*, 1529–1543. [CrossRef]
11. Pappalardo, G.; Mineo, S.; Caliò, D.; Bognandi, A. Evaluation of Natural Stone Weathering in Heritage Building by Infrared Thermography. *Heritage* **2022**, *5*, 2594–2614. [CrossRef]
12. Keppert, M. *Damage of Porous Stones by Salt Crystallization*; Čáchová, M., Fořt, J., Koňáková, D., Pavlík, Z., Černý, R., Eds.; WIT Transactions on Modelling and Simulation; WIT Press: Opatija, Croatia, 2015; pp. 333–341. [CrossRef]
13. Germinario, L.; Oguchi, C.T. Underground Salt Weathering of Heritage Stone: Lithological and Environmental Constraints on the Formation of Sulfate Efflorescences and Crusts. *J. Cult. Herit.* **2021**, *49*, 85–93. [CrossRef]
14. Germinario, L.; Andriani, G.F.; Laviano, R. Decay of Calcareous Building Stone under the Combined Action of Thermoclastism and Cryoclastism: A Laboratory Simulation. *Constr. Build. Mater.* **2015**, *75*, 385–394. [CrossRef]
15. Germinario, L.; Oguchi, C.T. Gypsum, Mirabilite, and Thenardite Efflorescences of Tuff Stone in the Underground Environment. *Environ. Earth Sci.* **2022**, *81*, 242. [CrossRef]
16. Ulusay, R. (Ed.) *The Complete ISRM Suggested Methods for Rock Characterization, Testing and Monitoring: 1974–2006*; Commission on Testing Methods; International Society for Rock Mechanics: Ankara, Turkey, 2007.
17. *EN1936:2006*; European Standard 1936: 2006 Natural Stone Test Methods—Determination of Real Density and Apparent Density, and of Total and Open Porosity. European Union: Maastricht, The Netherlands, 2006.
18. Siegesmund, S.; Durrast, H. Physical and Mechanical Properties of Rocks. In *Stone in Architecture. Properties, Durability*, 5th ed.; Siegesmund, S., Snethlage, R., Eds.; Springer: Berlin, Germany, 2014; pp. 97–224.
19. Mineo, S.; Pappalardo, G. The Use of Infrared Thermography for Porosity Assessment of Intact Rock. *Rock Mech. Rock Eng.* **2016**, *49*, 3027–3039. [CrossRef]
20. Mineo, S.; Pappalardo, G. InfraRed Thermography Presented as an Innovative and Non-Destructive Solution to Quantify Rock Porosity in Laboratory. *Int. J. Rock Mech. Min. Sci.* **2019**, *115*, 99–110. [CrossRef]
21. Mineo, S.; Pappalardo, G. Nondestructive Rock Porosity Estimation by InfraRed Thermography Applied to Natural Stones. *Constr. Build. Mater.* **2022**, *342*, 127950. [CrossRef]
22. Mineo, S.; Pappalardo, G. Preliminary Results on the Estimation of Porosity in Intact Rock through InfraRed Thermography. *ROL* **2016**, *41*, 317–320. [CrossRef]
23. Germinario, L.; Török, Á. Variability of Technical Properties and Durability in Volcanic Tuffs from the Same Quarry Region—Examples from Northern Hungary. *Eng. Geol.* **2019**, *262*, 105319. [CrossRef]
24. Germinario, L.; Török, Á. Surface Weathering of Tuffs: Compositional and Microstructural Changes in the Building Stones of the Medieval Castles of Hungary. *Minerals* **2020**, *10*, 376. [CrossRef]
25. Lukács, R.; Harangi, S.; Guillong, M.; Bachmann, O.; Fodor, L.; Buret, Y.; Dunkl, I.; Sliwinski, J.; Von Quadt, A.; Peytcheva, I.; et al. Early to Mid-Miocene Syn-Extensional Massive Silicic Volcanism in the Pannonian Basin (East-Central Europe): Eruption Chronology, Correlation Potential and Geodynamic Implications. *Earth-Sci. Rev.* **2018**, *179*, 1–19. [CrossRef]
26. DeWitt, D.P.; Nutter, G.D. *Theory and Practice of Radiation Thermometry*; Wiley: New York, NY, USA, 1988.
27. Thomachot-Schneider, C.; Gommeaux, M.; Lelarge, N.; Conreux, A.; Mouhoubi, K.; Bodnar, J.-L.; Vázquez, P. Relationship between Na_2SO_4 Concentration and Thermal Response of Reconstituted Stone in the Laboratory and on Site. *Environ. Earth Sci.* **2016**, *75*, 762. [CrossRef]
28. Thomachot-Schneider, C.; Vázquez, P.; Gommeaux, M.; Lelarge, N.; Conreux, A.; Drothière, X.; Mouhoubi, K.; Bodnar, J.-L. Thermal Response of Building Stones Contaminated with Salts. *Constr. Build. Mater.* **2019**, *226*, 331–344. [CrossRef]
29. Vazquez, P.; Thomachot-Schneider, C. Infrared Thermography as a Tool to Detect Increasing Cracking in Granitic Stones Exposed to High Temperatures. *J. Cult. Herit.* **2023**, *59*, 163–170. [CrossRef]
30. Franzosi, F.; Casiraghi, S.; Colombo, R.; Crippa, C.; Agliardi, F. Quantitative Evaluation of the Fracturing State of Crystalline Rocks Using Infrared Thermography. *Rock Mech. Rock Eng.* **2023**. [CrossRef]
31. Caliò, D.; Mineo, S.; Pappalardo, G. Digital Rock Mass Analysis for the Evaluation of Rockfall Magnitude at Poorly Accessible Cliffs. *Remote Sens.* **2023**, *15*, 1515. [CrossRef]

32. Mineo, S.; Caliò, D.; Pappalardo, G. UAV-Based Photogrammetry and Infrared Thermography Applied to Rock Mass Survey for Geomechanical Purposes. *Remote Sens.* **2022**, *14*, 473. [CrossRef]
33. Frodella, W.; Elashvili, M.; Spizzichino, D.; Gigli, G.; Adikashvili, L.; Vacheishvili, N.; Kirkitadze, G.; Nadaraia, A.; Margottini, C.; Casagli, N. Combining InfraRed Thermography and UAV Digital Photogrammetry for the Protection and Conservation of Rupestrian Cultural Heritage Sites in Georgia: A Methodological Application. *Remote Sens.* **2020**, *12*, 892. [CrossRef]
34. Pappalardo, G.; Mineo, S.; Carbone, S.; Monaco, C.; Catalano, D.; Signorello, G. Preliminary Recognition of Geohazards at the Natural Reserve "Lachea Islet and Cyclop Rocks" (Southern Italy). *Sustainability* **2021**, *13*, 1082. [CrossRef]
35. Guerin, A.; Jaboyedoff, M.; Collins, B.D.; Derron, M.-H.; Stock, G.M.; Matasci, B.; Boesiger, M.; Lefeuvre, C.; Podladchikov, Y.Y. Detection of Rock Bridges by Infrared Thermal Imaging and Modeling. *Sci. Rep.* **2019**, *9*, 13138. [CrossRef]
36. Grechi, G.; Fiorucci, M.; Marmoni, G.M.; Martino, S. 3D Thermal Monitoring of Jointed Rock Masses through Infrared Thermography and Photogrammetry. *Remote Sens.* **2021**, *13*, 957. [CrossRef]
37. Mineo, S.; Pappalardo, G. Rock Emissivity Measurement for Infrared Thermography Engineering Geological Applications. *Appl. Sci.* **2021**, *11*, 3773. [CrossRef]
38. Hall, C.; Hamilton, A. Porosities of Building Limestones: Using the Solid Density to Assess Data Quality. *Mater. Struct.* **2016**, *49*, 3969–3979. [CrossRef]

minerals

MDPI

Article

Unveiling the Secrets of an Artwork through Non-Invasive Investigations—Case Study of a 19th-Century Female Portrait

Adrian Rauca [1], Luminița Ghervase [2,*], Antonia Berdie [1] and Matei Agachi [1]

1 Conservation and Restoration Department, University of Art and Design Cluj-Napoca, 31 Unirii Street, 400098 Cluj-Napoca, Romania; adrian.rauca@uad.ro (A.R.); antonia.berdie.cr@uad.ro (A.B.); matei.agachi@uad.ro (M.A.)

2 National Institute of Research and Development for Optoelectronics, 409 Atomiștilor Street, 077125 Măgurele, Romania

* Correspondence: ghervase@inoe.ro

Abstract: This article aims to present the results of the investigations performed on a 19th-century oil painting on canvas belonging to the Conservation and Restoration Department of the University of Art and Design in Cluj-Napoca, Romania. The artwork depicting the portrait of a lady originating from an important Irish noble family (Judith Bunbury, 1785–1861) has been investigated using only non-invasive methods. The investigation protocol included digital photography in different light sources at different wavelengths, which was used to document the current condition of the painting, UV fluorescence, which highlighted the previous improper retouches and the presence of a varnish coat, Infrared reflectography, which brought to light the underdrawing made by the artist and X-ray Fluorescence (XRF) spectroscopy, an elemental analysis technique which indicated the chemical composition of the pigments, suggesting the use of lead white, a barium white pigment, yellow and red ochre, vermilion, cobalt blue, and a manganese-based brown pigment. The results of this entirely non-invasive investigation approach helped in choosing the most appropriate conservation and restoration methodology for the artwork.

Keywords: conservation and restoration; scientific investigation; non-invasive investigation; artwork; painting

1. Introduction

The process of investigating an artwork is a very important preliminary step made to reveal precious information about the technique, materials, and conservation state of the painting. The results of the investigations are useful to determine the manufacturing technique and the color palette of an artist and can also be useful for determining the proper conservation and restoration methodology.

After determining the materials used in creating the painting, the main aim throughout the conservation and restoration process is the use of compatible, preferably historically accurate, materials, which in time under different microclimate conditions will tend to age and react similarly to the original ones constituting the painting. In this way, the creation and amplification of additional physical tensions inside the layers is avoided.

The subject of this study is an oil painting on canvas from Ireland and created by an anonymous artist, most likely sometime in the middle of the 19th century (no signature or inscriptions were found). The artwork was acquired in 2022 from a British auction house by a private Romanian art collector and brought into the Conservation and Restoration Department of the University of Art and Design in Cluj-Napoca to be restored. Due to the poor state of conservation of the painting, a non-invasive investigation protocol has been performed. An inscription on the back of the painting suggests that it represents the Portrait of Judith Bunbury, born in 1782 in Ireland and died in 1861 in Cheltenham, England, and wife of Hans Allen, a Royal Irish Artillery colonel (as indicated by the "R.A."

inscription on the back of the painting). The portrait (genre-portraiture) belongs to the Early Victorian period, it is inspired by realism and it depicts an ethical essence because it shows the person depicted with a written text under the arm [1].

Based on the research done by J. Ferran [2], Judith was Capt. Benjamin Bunbury and Jane Hall's daughter and had seven children with Hans Allen: Jane Allen, John Tyd Allen, Sussana Allen, John Bunbury Allen, Hans Wallace Allen, Harriet Allen, and Elizabeth Allen [2]. Otherwise, J. Ferran lists her birth date as around 1785, which contradicts the inscription from the back of the painting.

A scrap of paper was found attached on the verso of the painting with the following text: "For Hans Rathborne from aunt Hal/r...". Based on the family tree (Figure 1), Judith's daughter Elizabeth Allen together with her husband, Capt. William Humphrey Rathborne, had two sons named Hans Rathborne: col. William Hans Rathborne (1841–1913) and lt. col. Hans Robert Rathborne (1843–1927). By the Victorian era traditions, the first name, and essentially the one used in society, of the firstborn male child (in this case William) was named after the father's father, while the second male child was named after the mother's father. Therefore, we can conclude that Col. Hans Robert was the one who received the painting as a gift probably from Aunt Harriet, born in 1818. The Bunbury family dates back to the 12th century, and it was formed in the Kingdom of England but had Norman origins.

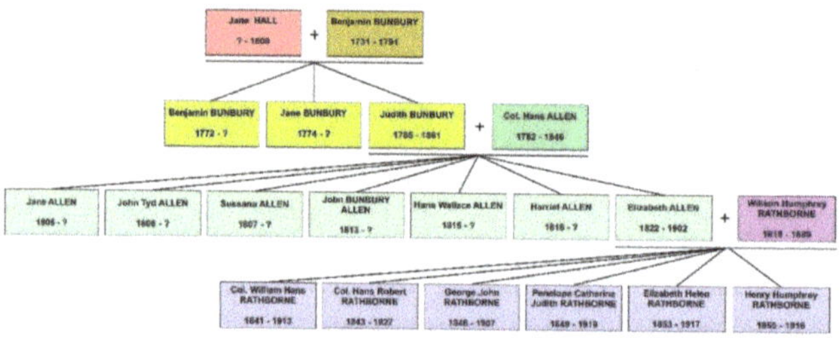

Figure 1. The Bunbury family tree.

We thus understand that the figure represented in the painting is descended from this historical family and due to the social and economic status of the family maintained over time, we assume that the artwork was painted by a significant painter. The quality of the painting technique states the same fact.

The study aimed to gather information about the materials and technique of the artist, in a completely non-invasive manner, in view of designing the optimal restoration process. Another goal of the paper was to establish a time frame for the painting, by using the information gathered from the investigation protocol, correlated with the available written historical information and research regarding the fashion attire of the sitting figure.

2. Materials and Methods

Before the conservation and restoration process of artwork, investigations and written/visual documentation are carried out to create a record of the painting's initial condition and characteristics. Nowadays, it has become common practice to set up multi-analytical non-invasive investigation protocols, to reduce as much as possible the invasive methods of investigation. For the painting under discussion, to minimize any potential harm to the studied object, which was in a poor state of conservation, all the investigations conducted were non-invasive. The original canvas of the artwork was very fragile due to the improper microclimate in which it was held before it entered the conservation and restoration studio. It was perforated in eight spots of which seven have been improperly "consolidated" with thick layers of overpainting on the front (more probably oil paint) and unsuitable cotton fabric patches with glue on the back. The paint layer adhered strongly to the canvas and only some small areas were flaking. Chromatic (yellowing and darkening) and textural (craquelures) alterations of the varnish layer took place due to aging and oxidation. The investigation protocol was designed to accomplish the main goals of the restorers: to determine the most appropriate way of treatment, to obtain the best results possible after the treatment, and to thoroughly document all steps of this process.

A Nikon D700 camera (12.1 megapixels and 35 mm full-frame sensor with AF-S NIKKOR 24–70 mm lens) was used to record high-resolution photographs of the painting under different illumination conditions. The spatial resolution at the acquisition was 4256 × 2832 pixels and the objective was set at f/3.5 diaphragm aperture and 50 mm focal distance. First, the artwork was examined in natural illumination. The uniformly scattered lighting was provided naturally, coming through the windows. This investigation was done right after the artwork entered the doors of the studio and helped us better understand its current condition. Second, the artwork was observed in transillumination, by pointing a powerful source of light (35 W, 98 daylight LED, 5000 °K) on the back of the painting (at 60–70 cm distance), to bring to light the inner features of the painting. This was followed by the examination of the painting in raking light, which highlighted the topography of the artwork, including uneven tensions of the support, degradation due to previous interventions, and density of the brush strokes. This non-damaging process consisted of placing the same source of light on the side of the painting to accentuate and reveal surface details such as texture, brushstrokes, and irregularities that may not be distinguishable under frontal lighting conditions. By positioning the source of light at a low angle (between 0 and 20°) to the surface of the artwork, shadows that emphasize the irregularities of the surface are created. The artwork can be examined from different angles by moving the light source around the studied object. Finally, ultraviolet (UV) fluorescence images were recorded, by placing an ultraviolet lamp in front of the artwork. In this way, the materials found on the surface of the artwork undergo temporary modifications called fluorescence. This phenomenon makes them easier to identify because each material has a distinguishable fluorescent property. The examination was done using the Luminaire Cts Art Lux 100 LV portable lamp with 8 UV LED lights, 365 nm wavelength. To collect only the fluorescent radiation, a 77 mm Hoya UV-blocking filter for the camera lens was used.

Infrared reflectography (IR) is a non-invasive technique of linear optical imaging that consists of producing an infrared image between 1000 and 1100 nm electromagnetic radiation of the investigated artwork. This technique reveals features underlying the pictorial layer, such as underdrawings, signatures, compositional and hidden pictorial changes (pentimento) made by the artist, and sometimes traces of subsequent overpainting [3]. The infrared photographs revealing the underdrawing of the painting were taken using Cts's System for reflectography Mod. Mir 10 New, which is composed of a MICRO IR10.2 color camera (6 Mpixel resolution) for infrared reflectography with a sensor sensitive to the infrared part of the spectrum, a 12.5 mm f1/8 lens (RICOH TV Lens), and an infrared filter (IR1100). The illumination sources were two common incandescent lamps directed from the sides at approximately 5 m distance, radiating enough infrared rays for our purpose.

The microscopic investigations were performed directly on the painting using the portable video microscope (CTS Videomicroscope I-Scope USB); for the data collection, Moritex Viewer 2.2 was used. Thanks to its integrated LED illumination system and capacity of magnifying the image up to 200×, a close observation of the artwork has been provided and important characteristics have been revealed such as textures and degradation phenomena. In the investigation process, 50× (FOV 4.5 mm) and 200× (FOV 1 mm) magnifying lenses were used.

The elemental composition of the inorganic pigments was determined using the X-ray Fluorescence spectrometry method. This analysis was performed using a portable handheld energy-dispersive spectrometer Bruker TRACER III-SD. The equipment was operated in hand-held mode, at ~2 mm distance from the painting surface, at 40 kV tube voltage, 11 μA current intensity, for 5 s analysis time, without filtering of the beam. Identification of elements was performed with ARTAX software (Bruker, AXS GmbH, Mannheim, Germany) using standard Bayesian deconvolution, allowing a semi-quantitative approach. Data plotting was done using OriginPro 2021 (OriginLab Corporation, Northampton, MA, USA).

The main advantage of using this portable instrument is that it provides a non-invasive, non-destructive analysis of pigments and inorganic materials in artworks.

All the investigations mentioned above represent a very important step before the restoration process because they are meant to reveal important information about the materials that have been used by the artist and about its technique, helping the conservator understand the artwork to approach suitable treatments.

3. Results and Discussion

3.1. *Art Historical Analysis*

The 63.5 × 76 cm painting depicts the bust portrait of an elderly lady with her right hand raised near the chest, holding a golden Victorian lorgnette (reading glasses with a handle). Her elbow rests upon an open book or some sheets of paper placed on a table, reflecting the typical conventions of the Victorian era. The figure is dressed in a black mourning ensemble, called "widow's weeds". Several elements, such as the strength of the clothing elements, jewelry, and also the fact that she is represented with glasses in her hand and a text under her elbow resting on a desk, point towards the idea that the woman represented in this painting was a well-educated, wealthy person.

Although the elderly lady is looking at the viewer with some semblance of warmth, she still conveys a sense of dignity and authority, rather than a friendly demeanor. This was common place in Victorian England, where society looked down on smiles and grins, which were associated with madness or lewdness.

The chromatic palette of the painting is represented by warm colors, especially brown and ochre of various intensities. The white color of the lace is warmed with ochre and the garment is close to black. The background color is also a lighter brown. The color of the face and hand is warm; the dominant color is yellow ochre, with different lighter or darker shades. The painter applied an intense red color for the coloring of the lips and more faded redness in the cheeks. It is obvious that the painting was realized in an enclosed space, with the light source illuminating the person probably being a lamp or candles.

Regarding the period in which the portrait was painted, some information can be deduced from written archive information. If Hans Robert Rathborne was born in 1843 (as mentioned in Burke's Peerage, a British genealogical publisher [4]), then the painting was commissioned at a later date. Considering that Judith died in 1861, and if the portrait was made using the real-life model, it could mean that the portrait was painted between 1843 and 1861. Nevertheless, if it was commissioned after her death (painted after an earlier painting or photograph) then it could have been commissioned before the last couple of decades of the 19th century. Although Aunt Harriet's time of death is unknown, considering she was born in 1818, her death probably occurred long before the turn of the century.

3.2. *Visual Analysis*

3.2.1. Examination in Natural Light

Generally, one of the first steps is to observe the painting in visible light [5]. The documentation of the artwork's state of conservation from the moment it enters the studio is an important step in the restoration process. One way to document its condition is by taking high-resolution photographs in good lighting. When photographing the artwork in direct lighting, the light falls uniformly across the entire surface of the painting.

After photographing the artwork in visible light (Figure 2), it was noticed that the textile support is made out of a single piece of linen fabric that has been perforated in eight spots, mainly in the background area. Of these, seven perforations have been improperly consolidated with a thick layer of oil paint (overpaint) on the front and with unsuitable patches on the back (indicated in Figure 2b by yellow dashed boxes). The oil content of the overpaint has been proven by performing an oil test on an isolated area using a saturated solution of Sudan Black B. The patches were mechanically unsuitable due to the animal glue used to adhere them to the holes and tears. The fact that the adhesive is protein-based (animal glue) was proven by a protein test performed on a patch using an acid fuchsin reagent. The glue, which lost its elasticity and contracted over time, caused some aesthetically disturbing bulges on the front of the painting and distinctive craquelure in the paint layer. The fact that one of the perforations has not been consolidated indicates that it might be more recent than the others. Additionally, it could be noticed that both the paint layer and the varnish layer were evenly applied. The paint layer has strong adhesion to the preparation layer, with only a few areas on the shoulder and above the right hand in which lacunae and flaking were seen (Figure 3). Due to aging, chromatic (yellowing and darkening) and textural (craquelures) alterations of the varnish layer took place. The chromatic alterations are probably photo-induced due to the fact that the edges of the painting layer that were covered by the frame did not show as severe an alteration as the areas exposed to light and oxygen.

(a) (b)

Figure 2. The portrait of Judith Bunbury—photographs of the front (**a**) and back (**b**) of the painting in visible light.

Figure 3. Flaking and lacunae of painting layer, photographed in visible light.

3.2.2. Transillumination

This type of investigation consists of placing a source of light on the back of the painting. It is commonly referred to as "backlighting". Shining the light through an object helps analyze its internal features and reveals details that may not be visible through other examination methods.

This investigation method proved to be very effective in highlighting the damages and imperfections of the canvas, technical aspects, and previous interventions. The most visible flaws seen under this type of investigation were the perforations and the improper patching done on the back using thick cotton fabric, as can be seen in Figure 4a,b. Although patching was commonly used for lining and locally stabilizing gaps in paintings, their presence on the canvas can induce unwanted effects, such as contraction or swelling. The animal glue, which lost its elasticity, became brittle and contracted over time, causing bulges on the front of the painting and deformations of the textile support. Furthermore, it caused the appearance of distinctive craquelure in the paint layer in these areas [6].

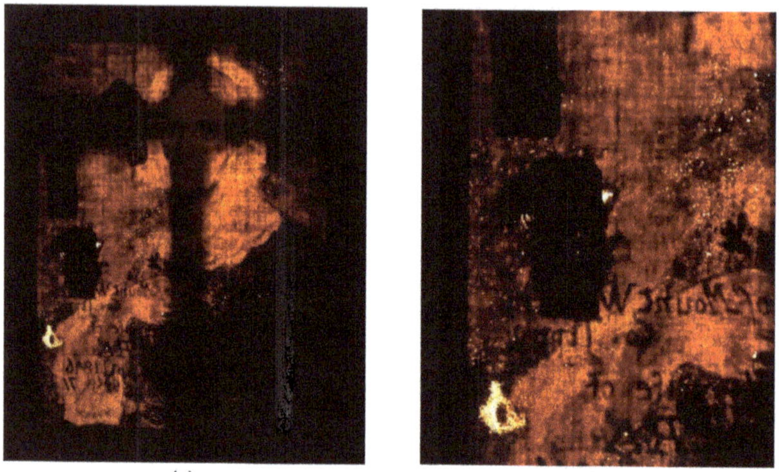

(**a**) (**b**)

Figure 4. Photography with transilluminating light; overview (**a**) and detail (**b**).

Transillumination also revealed details about the painting technique, such as the fact that some areas were painted in light coats of paint and others were applied in a pasty manner. Backlighting is also important to see if the added paint exceeds the margins of the existing lacunae. In this case, it showed that the overpainting was not limited to the edges of the lacunae, covering large areas of the original painting layer as well.

3.2.3. Observation in Raking Light

Raking light is a non-destructive investigation method used to examine the surface of different types of artworks such as wall, panel, or canvas paintings. This type of investigation highlights the tridimensional aspect of the surface of the artwork providing information regarding the technique and conservation state. It allows conservators to see the painting from different angles and to observe how the light interacts with its surface.

As can be seen in Figure 5a,b, the examination of the painting in raking light highlighted the thick layers of oil paint used to improperly retouch the painting and the deformations and bulges of the canvas caused by the patches on the back (alterations highlighted by the yellow dashed boxes). The areas painted in a pasty manner are also easily distinguished under the radiant source of lighting.

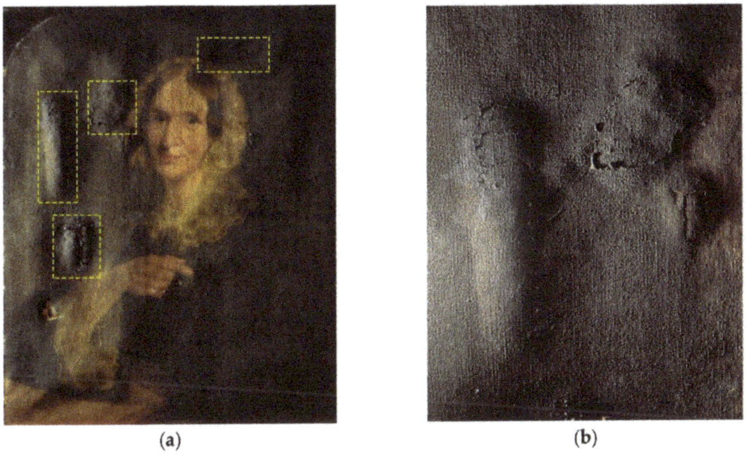

Figure 5. Photography with raking light; overview (a) and detail (b).

3.3. Imaging Analysis

3.3.1. Videomicroscopy

Videomicroscopy is a non-invasive investigation technique that does not imply sampling and provides a close in situ observation of the artwork highlighting the main characteristics of the technique and materials used by the artist and also the structural damages [7].

Thanks to its incorporated lighting system, the I-scope equipment generated detailed images of the examined elements. Therefore, the I-scope provided a close view of the canvas highlighting its texture and its warp and weft (Figure 6a), the varnish coat and its patinated, agglutinated look caused by aging and light/sun exposure (Figure 6b), of the improperly made retouches by a previous restorer (Figure 7a), and of the flaking paint layer (Figure 7b). In most areas, the paint layers separated, while the ground layer remained adhered to the textile support. This might have been caused due to loss of adhesion between the ground and paint layers. If the ground layer was applied early on to the canvas, but the painting took place sometime later, the excessive drying of the ground layer (containing lead, which has a drying effect) and an insufficient quantity of oil binder in the subsequent paint layer could have caused this form of flaking. Another explanation could be damage caused by humidity. The exchange of moisture between the front layers and the textile support of the painting occurs through the craquelure, and thus, moisture may accumulate between the paint and ground layer. The moisture may have softened the glue and ground layer, while the paint layers remained inelastic. This may have led to the lifting and separation of the layers.

Figure 6. The texture of the canvas (a) and of the varnish coat (b).

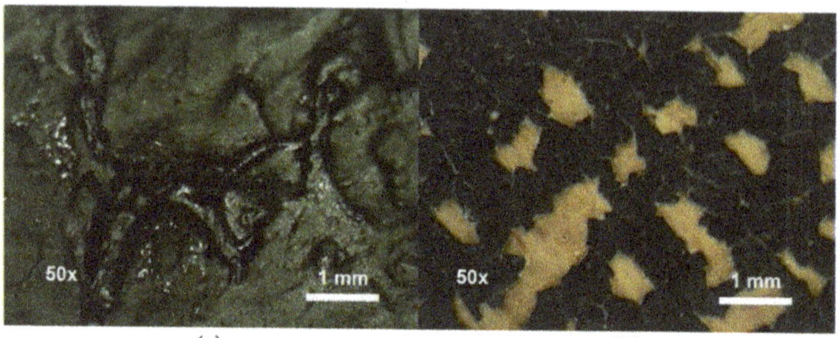

(a) (b)

Figure 7. Defects on the surface: improperly made retouch (**a**) and flaking (**b**).

3.3.2. UV Fluorescence Imaging

Even if the varnish of a painting can be easily observed in good lighting, using UV fluorescence helps to determine how homogenous the application of the varnish layer is [8]. It also allows us to see where retouching was done using similar colors. The varnish of a painting provides a protective role for the painting acting as a barrier to dirt, smoke, and injury caused by possible mechanical scratches. It also holds an aesthetic value providing depth and clarity to the colors [9]. Furthermore, the varnish layer protects against atmospheric pollutants, such as chlorine or sulfur emissions that can alter the color of certain pigments. It also offers protection against ultraviolet rays (which can cause the oxidation of the oil-based binder, as well as the discoloration of some organic dyes or the alteration of some inorganic pigments) and infrared rays (which, by heating the substrate, can cause the loss of mechanical resistance within the layers), originating from natural and/or artificial light sources. Putting the artwork under this method of investigation revealed that the painting has an evenly applied layer of varnish and some retouches that are improperly made (highlighted in Figure 8 by the yellow dashed boxes). The darker areas that can be seen in Figure 8, which can be easily distinguished from the original paint, correspond to areas where retouching has been made. Also, the greenish tint under UV light indicates the use of a natural resin varnish coat. The retouches were made in a single layer of paint, meaning they were applied during a single restoration intervention when the patches were also applied. The retouches did not cause any color change in the proximity of the retouched areas.

(a) (b)

Figure 8. Photography in ultraviolet light; overview (**a**) and detail (**b**).

3.3.3. Infrared Reflectography (IR)

Infrared reflectography is a non-invasive optical technique that consists of producing an infrared image of the investigated artwork that reveals features underlying the pictorial layer. Infrared reflectography is one of the routine methods of examining paintings since 1950 and it is the tool of choice for researching the underdrawings. The underdrawing is the drawing that the artist made on the already prepared canvas for painting, it was then covered by subsequent paint layers. Depending on their chemical composition, certain materials absorb the IR radiation (they appear dark), while others reflect it, meaning that they appear brighter, thus the underdrawings, compositional changes, signatures, or underlying brushstrokes may become visible. Usually, iron, aluminum, manganese, barium, copper, cobalt, and graphite-based pigments and materials (for example earth pigments, ochres, cobalt blue, malachite, and azurite) absorb the radiation and usually appear dark. Cadmium, zinc, chromium, and titan-based pigments usually transmit infrared radiation and reflect it quite well (for example, cadmium yellows and reds, and zinc and titan white), thus appearing bright. Lead, tin, mercury, and sulfur-based pigments strongly reflect the radiation and appear very bright (for example lead white, vermillion, and ultramarine). Of course, the reflection of the infrared radiation depends on the thickness of the paint layer as well as on the quantitative composition of the various elements. While these features are not conclusive, they can be used as a starting point along with other analytical investigations in identifying some of the pigments used.

Sometimes superimposition of the IR images occurs when both the paint layers and the underdrawings can be seen, thus making it more difficult to interpret the images [10].

The use of IR allowed the underdrawing of the painting to be seen, which would not have been possible without the naked eye. The underdrawing made in graphite was obvious under the portrait, under the bonnet, under the neck area, and under the hand (Figure 9).

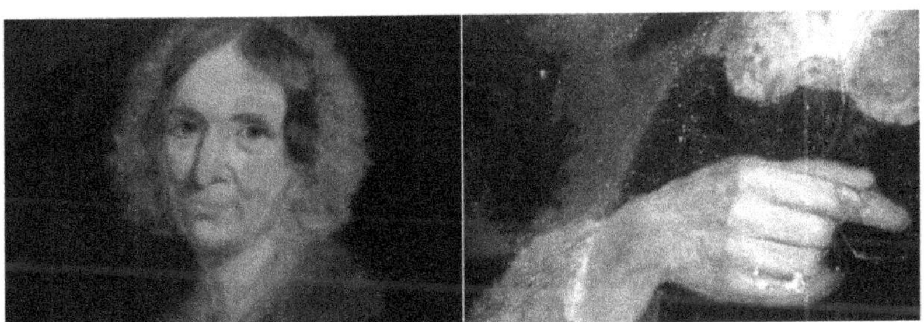

(a) (b)

Figure 9. Infrared photographs showing the underdrawings made in graphite: portrait area (**a**); right-hand area (**b**).

There was not any signature of the painter found on the painting. This fact may be due to several reasons. One reason could be that the painter signed the painting on an additional piece of paper attached to the canvas. Another reason, likely the most plausible one, is related to the fact that the studied painting is a commissioned work. In the Victorian era, some artworks were left unsigned because they were not for sale in front of an audience, they were not showcased to the public at soirees or in great halls. As it is in this case, the painting was a commissioned work probably by Aunt Harriet, who offered it as an exclusive gift to Hans Robert Rathborne, in memory of his grandmother, Judith Bunbury. Portraits received as gifts often sat in a personal library or a private study, meaning they were not intended for the public eye, but rather for a private audience.

3.4. Elemental Analysis

X-ray Fluorescence (XRF) spectrometry is a non-destructive method of investigation that allows obtaining an accurate and quick identification of inorganic materials and has been widely used in the study of art objects in past decades [11]. XRF studies have been performed on different types of artworks such as manuscripts, wall paintings, wood panels, and canvas paintings [12–16]. For the aim of this study, the investigation was carried out on 10 distinct analysis zones, covering the entire color range, as can be seen in Figure 10.

Figure 10. Visible light photography with the points analyzed with XRF (numbers represent analysis areas).

Figure 11 shows the XRF spectra for some representative areas: blue (area 3), white (area 5), scarlet (area 6), and brown (area 9). As can be seen from the figure, the acquired spectra were dominated by the high lead (Pb) lines, present in all analyzed areas. Lead has been used since ancient times to produce variations of white, red, or yellow pigments [11,17]. However, XRF alone cannot differentiate between different pigments with similar elemental composition. Moreover, it cannot determine where the lead signal is coming from [11], and given the fact that lead was found in all analyzed areas, the signal may be coming from an underlayer [18,19]. Also, based on the fact that the painting in question is an oil painting, it could have been used as a drier for the binder [20]. So, the presence of the lead lines could be due to one or more of the previously mentioned reasons.

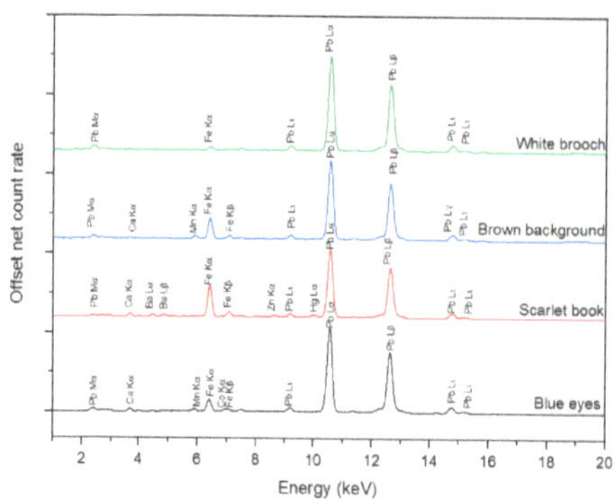

Figure 11. XRF spectra of selected areas.

The fact that, aside from the lead lines, no other significant elements were detected in the spectrum of the white area (5), suggests that lead white could have been used as a white pigment.

In addition, similar lead intensities were found for area 8 (lacuna in the background), which sustains the idea that lead was used in the ground layer as well. Besides lead, calcium (Ca) was another element identified in every analyzed area. The use of lead white mixed with calcium carbonate in primers was frequently used until the 20th century [21] because of its qualities: it offers good and luminous coverage in oil, has a drying effect on the oil medium, and also provides a low oil absorption, thus counteracting the uneven absorbency of the overlapping paint layers in time.

This information was crucial in further restoration interventions, namely at the filling of lacunae, where the choice of the filler would consist of calcium carbonate with a white pigment mixed with animal glue.

For the background (areas 8 and 9), the manganese line (Mn Kα) could be seen, which points towards the use of brown manganese-based pigment.

The spectrum corresponding to area 6 shows both slightly higher iron (Fe) lines, indicating the use of an iron-based pigment, and traces of mercury (Hg), as can be seen in Figure 10 (scarlet book). Although the high lead lines made it difficult to assess the presence of sulfur, which appears in the same energy range as the Pb Mα lines, the occurrence of mercury traces indicates the use of the mercuric sulfide pigment (HgS). This is commonly referred to as cinnabar or its synthetic variant, vermilion [17], a pigment that is known to degrade with time, leading to the darkening of its original color. In Europe, both natural (from the cinnabar mineral) and synthetic vermilion were used simultaneously [22] until the end of the 19th century.

Another interesting aspect is the fact that traces of barium (Ba) and zinc (Zn) were evident in the spectrum of area 6 but also appeared at lower levels in some of the other areas.

The presence of Pb, along with Zn and Ba indicates the artist's technique of combining multiple white pigments to obtain the desired tones. The combination of zinc and barium could come from lithopone, a mixture of zinc sulfide and barium sulfate ($ZnS{\cdot}BaSO_4$), introduced in the second half of the 19th century [17,23], but it could also point towards the use of zinc white, which is zinc oxide, and barium sulfate, known as *Blanc Fixe* [24,25].

XRF analysis also revealed the presence of cobalt traces in the spectrum recorded from the area of the eye (3). While only traces of cobalt could be seen, correlated with the visual aspect, it can infer the use of a blue cobalt-based pigment. XRF spectroscopy alone cannot directly indicate which type of pigment could have been used, as several blue cobalt-based pigments were in use during the 19th century. However, the absence of certain elements, such as As, Sn, Zn, or Cr can rule out the possibility of certain pigments, including cerulean blue, dark cobalt blue, smalt, cobalt chromite, or cobalt spinel, thus leaving the most probable pigment as cobalt blue, which has been used since the beginning of the 19th century [26].

4. Conclusions

The investigation campaign of the artwork was a crucial preliminary step that unveiled valuable information about its complexity, conservation state, technique, previous interventions (such as improper retouches and patches), and chemical composition of its materials.

The examination in natural light played a pivotal role in this investigation protocol revealing information about the general state of the painting and its flaws. By shining light through the back of the artwork, hidden damage, weaknesses like perforations, technical aspects, and previous improper restorations, which have led to the structural deterioration of the painting, were highlighted. It also provided an insight view into the artist's working process. The artwork's topography and texture were highlighted thanks to the raking light investigation in which shadows were generated by the low-angle positioned source of light. This method of investigation allowed the conservators to appreciate and better understand the artistic techniques that were used in the process of creating this artwork, represented by a combination of light and pasty brushstrokes. The use of UV lighting proved to be a valuable investigation technique that made the improper previous retouches obvious, with several darker areas, which correspond to newer materials. Moreover, the use of a natural resin varnish was indicated by the greenish tint of the UV images. Videomicroscopy was useful in evidencing the uniformly applied varnish layer, which had an agglutinated look, with chromatic alterations due to light exposure. The underdrawings were highlighted with the help of IR reflectography, in the upper half and middle of the drawing, in the bonnet, neck, and hand areas. The XRF analysis revealed the chemical composition of the used inorganic pigments: white areas showed lead, along with barium and zinc, red areas showed both iron and mercury, the brown background showed the presence of a manganese-based pigment, while the blue was obtained with a cobalt-based pigment, and the preparation layer showed both calcium and lead.

In summary, the investigation protocol carried out on the artwork significantly enhanced the understanding of it by gathering information about its condition, previous interventions, technique, and chemical composition. Of course, although not available at the time of the investigation, the methodology would be complete by adding some portable, non-invasive spectroscopic techniques, such as Raman or Fourier-transform infrared spectroscopy, for the identification of the molecular composition of the varnish, the glue used for patching or the impregnation layers, if any. There is no debate that a molecular analysis would have completed the investigations, but overall, the results obtained were, in this case, sufficient for the start of the restoration process, as they have helped experts understand the painting's historical and cultural context to approach a targeted restoration methodology. For instance, for the filling of the lacunae, the filler will consist of calcium carbonate, and animal glue with added titanium white, to match as closely as possible to the appearance and properties of the original ground layer. Additionally, by identifying some of the pigments, the use of acids or bases during cleaning interventions has become safe as they will not cause any color alteration.

Furthermore, by establishing a connection between the materials used by the artist and the written information available about the Bunbury family, in correlation with the painting technique used and the fashion attire worn by the model, the team successfully narrowed down the time frame in which the artwork was painted: the portrait was commissioned between 1843 and possibly 1861.

Author Contributions: Conceptualization, A.R. and L.G.; methodology, A.R.; investigation, A.R., L.G. and A.B.; resources, A.R.; writing—original draft preparation, A.R., A.B. and M.A.; writing—review and editing, L.G. and A.R. All authors have read and agreed to the published version of the manuscript.

Funding: This work was carried out through the project UpUAD funded by the National Council for the Financing of Higher Education, grant number CNFIS-FDI-2023-F-0468, and through the Core Program within the National Research Development and Innovation Plan 2022–2027, project no. PN 23 05, and Program 1—Development of the national research-development system, Subprogram 1.2—Institutional performance—Projects to finance the excellent RDI, Contract no. 18PFE/30.12.2021, carried out with the support of MCID.

Data Availability Statement: Data are available from the corresponding author upon reasonable request.

Acknowledgments: Not applicable.

Conflicts of Interest: The authors declare no conflict of interest.

References

1. Crowther, P. Awakening Beauty. The Crowther-Oblack Collection of Victorian Art. Available online: https://www.victorianweb.org/painting/awakeningbeauty/style.html (accessed on 10 June 2023).
2. Ferran, J. Legacy 9.0. Family Tree. Available online: https://www.monchique.com/Ochanoff/ohanov/ochanoff/7001.htm (accessed on 10 June 2023).

3. Amat, A.; Miliani, C.; Brunetti, B.G. Non-Invasive Multi-Technique Investigation of Artworks: A New Tool for on-the-Spot Data Documentation and Analysis. *J. Cult. Herit.* **2013**, *14*, 23–30. [CrossRef]
4. Burke, B. *A Genealogical and Heraldic History of the Landed Gentry of Ireland*; Harrison & Sons: London, UK, 1912.
5. Van Aspren de Boer, J.R.J. An Introduction to the Scientific Examination of Paintings. In *Scientific Examination of Early Netherlandish Painting*; Aspren De Boer, J.R.J., Filedt Kok, J.P., Dishoeck, F., Eds.; Brill: Leiden, The Netherlands, 1975.
6. Teixeira, A.C. Canvas Support Impregnation Materials and Techniques: A Study of Portuguese Painting and Its Conservation Issues. *CeROArt* **2016**. [CrossRef]
7. Mureșan, T. Videomicroscopia Portabilă În Analiza Picturii Murale. *Caiet. Restaurării* **2013**, 170–177.
8. Rauca-Bencze, F. Metodologii Istorice de Dublare a Picturilor În Ulei Pe Suport Textil. Studiu de Caz- Pelerinii Din Emaus Din Colecția Muzeului de Artă Cluj-Napoca. *Caiet. Restaurării* **2019**, *1*, 208–223.
9. Vandivere, A.; Wadum, J.; van den Berg, K.J.; van Loon, A. From 'Vermeer Illuminated' to 'The Girl in the Spotlight': Approaches and Methodologies for the Scientific (Re-)Examination of Vermeer's Girl with a Pearl Earring. *Herit. Sci.* **2019**, *7*, 66. [CrossRef]
10. Pinna, D.; Galeotti, M.; Mazzeo, R. Scientific Examination for the Investigation of Paintings: A Handbook for Conservator-Restorers. In *Scientific Examination for the Investigation of Painting: A Handbook for Conservator-Restorers*; Centro Di: Florence, Italy, 2009; ISBN 978-88-7038-474-1.
11. Shugar, N.A.; Mass, L.J. *Handheld XRF for Art and Archaeology*; Leuven University Press: Leuven, Belgium, 2012.
12. Cucci, C.; Bracci, S.; Casini, A.; Innocenti, S.; Picollo, M.; Stefani, L.; Rao, I.G.; Scudieri, M. The Illuminated Manuscript Corale 43 and Its Attribution to Beato Angelico: Non-Invasive Analysis by FORS, XRF and Hyperspectral Imaging Techniques. *Microchem. J.* **2018**, *138*, 45–57. [CrossRef]
13. Antunes, V.; Valadas, S.; Serrão, V.; Carvalho, M.L.; Candeias, A.; Mirão, J.; Cardoso, A.; Pessanha, S.; Manso, M. Josefa d' Óbidos Workshop from Panel to Canvas. Multianalytical Approach to Materials and Technical Evolution of the Most Significant Portuguese Painting Workshop of the 17 Th Century. *J. Mol. Struct.* **2019**, *1188*, 31–41. [CrossRef]
14. Cortea, I.M.; Ghervase, L.; Ratoiu, L.; Rădvan, R. Application of Spectroscopic and Hyperspectral Imaging Techniques for Rapid and Nondestructive Investigation of Jewish Ritual Parchment. *Front. Mater.* **2020**, *7*, 601339. [CrossRef]
15. Kokiasmenou, E.; Caliri, C.; Kantarelou, V.; Germanos Karydas, A.; Romano, F.P.; Brecoulaki, H. Macroscopic XRF Imaging in Unravelling Polychromy on Mycenaean Wall-Paintings from the Palace of Nestor at Pylos. *J. Archaeol. Sci. Rep.* **2020**, *29*, 102079. [CrossRef]
16. Klisińska-Kopacz, A.; Obarzanowski, M.; Frączek, P.; Moskal-del Hoyo, M.; Gargano, M.; Goslar, T.; Chmielewski, F.; Dudała, J.; del Hoyo-Meléndez, J.M. An Analytical Investigation of a Wooden Panel Painting Attributed to the Workshop of Lucas Cranach the Elder. *J. Cult. Herit.* **2022**, *55*, 185–194. [CrossRef]
17. Eastaugh, N.; Walsh, V.; Chaplin, T.; Siddall, R. *Pigment Compendium: A Dictionary and Optical Microscopy of Historical Pigments*; Eastaugh, N., Walsh, V., Chaplin, T., Siddall, R., Eds.; Elsevier: Amsterdam, The Netherlands, 2008.
18. Duran, A.; Siguenza, M.B.; Franquelo, M.L.; de Haro, M.C.J.; Justo, A.; Perez-Rodriguez, J.L. Murillo's Paintings Revealed by Spectroscopic Techniques and Dedicated Laboratory-Made Micro X-ray Diffraction. *Anal. Chim. Acta* **2010**, *671*, 1–8. [CrossRef] [PubMed]
19. Pięta, E.; Proniewicz, E.; Szmelter-Fausek, B.; Olszewska-Świetlik, J.; Proniewicz, L.M. Pigment Characterization of Important Golden Age Panel Paintings of the 17th Century. *Spectrochim. Acta Part A Mol. Biomol. Spectrosc.* **2015**, *136 Pt B*, 594–600. [CrossRef]
20. Laporte, L.; Ducouret, G.; Gobeaux, F.; Lesaine, A.; Hotton, C.; Bizien, T.; Michot, L.; de Viguerie, L. Rheo-SAXS Characterization of Lead-Treated Oils: Understanding the Influence of Lead Driers on Artistic Oil Paint's Flow Properties. *J. Colloid Interface Sci.* **2023**, *633*, 566–574. [CrossRef] [PubMed]
21. Stols-Witlox, M. Twentieth-Century Grounds. In *Conservation of Easel Paintings*; Stoner, J.H., Rushfield, R., Eds.; Routledge: London, UK, 2012; pp. 168–188.
22. Gettens, R.J.; Feller, R.L.; Chase, W.T. Vermilion and Cinnabar. In *Artists' Pigments. A Handbook of Their History and Characteristics, Volume 2*; Roy, A., Ed.; National Gallery of Art: Washington, DC, USA, 2012; pp. 159–182.
23. Macchia, A.; Schuberthan, L.M.; Ferro, D.; Colasanti, I.A.; Montorsi, S.; Biribicchi, C.; La Russa, M.F. Analytical Investigations of XIX–XX Century Paints: The Study of Two Vehicles from the Museum for Communications of Frankfurt. *Molecules* **2023**, *28*, 2197. [CrossRef] [PubMed]
24. Valentini, F.; De Angelis, S.; Marinelli, L.; Zaratti, C.; Colapietro, M.; Tarquini, O.; Macchia, A. Multianalytical Non-Invasive Characterization of 'Mater Boni Consilii' Iconography Oil Painting. *Heritage* **2023**, *6*, 3499–3513. [CrossRef]
25. Veiga, A.R. Oil Painting on Tinplate by Francisco José Resende. *CeROArt* **2010**. [CrossRef]
26. Gettens, R.J.; Stout, G.L. *Painting Materials: A Short Encyclopaedia*; Dover Publications, Inc.: New York, NY, USA, 1966.

minerals

MDPI

Article

Multi-Analytical Analysis of Decorative Color Plasters from the Thracian Tomb near Alexandrovo, Bulgaria

Georgi Avdeev [1], Rositsa Kukeva [2], Denitsa Yancheva [3], Valentin Mihailov [4], Vani Tankova [4], Momtchil Dimitrov [3], Georgi Nekhrizov [5], Radostina Stoyanova [2] and Bistra Stamboliyska [3,*]

1 Institute of Physical Chemistry, Bulgarian Academy of Sciences, Acad. G. Bonchev Str., Build. 11, 1113 Sofia, Bulgaria; g_avdeev@ipc.bas.bg

2 Institute of General and Inorganic Chemistry, Bulgarian Academy of Sciences, Acad. G. Bonchev Str., Build. 11, 1113 Sofia, Bulgaria; rositsakukeva@yahoo.com (R.K.); radstoy@svr.igic.bas.bg (R.S.)

3 Institute of Organic Chemistry with Centre of Phytochemistry, Bulgarian Academy of Sciences, Acad. G. Bonchev Str., Build. 9, 1113 Sofia, Bulgaria; denitsa.pantaleeva@orgchm.bas.bg (D.Y.); momtchil.dimitrov@orgchm.bas.bg (M.D.)

4 Institute of Solid State Physics, Bulgarian Academy of Sciences, 72 Tzarigradsko Chaussée, 1784 Sofia, Bulgaria; valentin@issp.bas.bg (V.M.); vtankova@issp.bas.bg (V.T.)

5 National Archaeological Institute with Museum, Bulgarian Academy of Sciences, 2 Saborna Str., 1000 Sofia, Bulgaria; nehrizov@gmail.com

* Correspondence: bistra.stamboliyska@orgchm.bas.bg

Abstract: In the present contribution, we report the results from a study on the ancient technology used to create decorative color plasters in the Thracian tomb near the village of Alexandrovo, Bulgaria. A series of fragments of red, black, grey, white and brown colored lime plasters from the dromos and funeral chamber were investigated by laser-induced breakdown spectroscopy, X-ray diffraction analysis, infrared spectroscopy, paramagnetic electron resonance spectroscopy and differential scanning calorimetry. Based on the combined analytical data, it was possible to identify the pigments, fillers and other materials in the composition of the decorative plasters in the interior, as well as to clarify the technological features related to the plaster creation. The results demonstrated that the murals were implemented on two layers in the case of white, black, grey and brown decoration—first coarse mortar, followed by a white, fine mortar, which usually was made of calcite. In the case of red decoration, a pigment was added to the fine mortar to achieve a colored surface. The pigments were identified as mostly traditional mineral pigments—calcite, kaolinite, red natural ochres (colored earth), brown colored earth and black pigment (amorphous C). The use of the fresco technique is implied by the major participation of calcite and the absence of organic binder in all of the painting layers.

Keywords: Thracian tomb; Alexandrovo village; Bulgaria; LIBS; XRD; EPR; IR; DSC

1. Introduction

The Thracian civilization reached its peak between the fifth and third centuries BC in present-day Bulgaria. Recent excavations have uncovered a series of tombs containing gold, silver, and bronze artifacts, providing evidence of the high level of culture achieved [1]. The unique tomb at the village of Alexandrovo was discovered accidentally at the end of 2000 in a large burial mound 1 km southeast of the village. During the subsequent rescue excavations, archaeologist Georgi Kitov and his team uncovered its entrance and explored and documented the tomb. The tomb dates back to the end of the fourth century BC [2]. It was found that it consists of a long corridor (dromos), a pre-burial chamber and a burial chamber. It is oriented in an east–west direction, with the entrance facing east. The corridor is made of different sized blocks in irregular rows and is covered with horizontally stacked stone slabs. The ante-burial chamber is quite small. It has a rectangular plan and is covered by a trapezoidal cantilevered vault. The circular burial chamber has a beehive-shaped dome made of segmental blocks, with the vaulting starting from the first row. It contained a burial bed that was found destroyed (Figure 1).

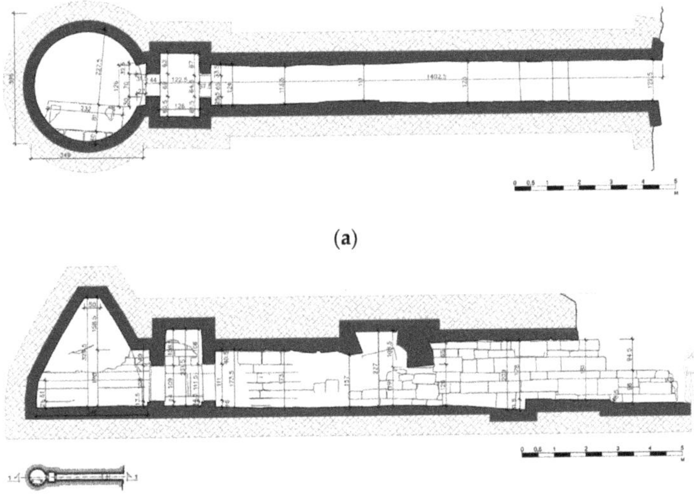

(**a**)

(**b**)

Figure 1. Plan 1:50 (**a**) and Section 1:50 (**b**) of the tomb at the village of Alexandrovo (author M. Kamenova, 2021).

The two chambers were constructed of precision-worked blocks with rusticated faces and were closed with double-leaf stone doors. The floors of the three chambers were covered with stone slabs, with a thick mortar applied to the floor in the burial chamber [3]. The tomb was constructed by blocks of rhyolite tuffs, widespread in the Eastern Rhodopes [4].

The walls of the chambers and the westernmost part of the dromos are covered with mortar plaster, on which the frescoed decoration of the tomb is located. It makes this monument one of the most significant representatives of Thracian tomb architecture with artistic decoration. On the north and south walls of the dromos, there are poorly preserved battle scenes between horsemen and foot warriors. The walls of the ante-burial chamber are covered with several parallel, horizontal bands, the first of which is white, the second one consists of black panels separated by vertical white bands, and the third one is ornamented with white spiral motifs on a dark background. The gabled roof is covered with a red-colored plaster and the horizontal part with white. The trapezoidal lintels above the entrances also depict battle scenes, although they have not been fully preserved (Figure 2).

The burial chamber is the most richly decorated. At the base is a deep red-shaped plinth, above which is a figural frieze depicting several human figures on a white background, interpreted as a funeral feast scene. Unfortunately, much of this frieze has been destroyed. Above the frieze follows a strip in black, of which a small part also survives, and above this a broad band colored red. On it, opposite to the entrance of the chamber, an inscription "ΚΟΞΙΜΑCΗC ΧΡΗCΤΟC" is scratched in two lines and depicts the face of a young man in profile on the left. Experts differ in their opinions about this unique display of individuality. Kitov suggests that it is the signature of the artist who painted the tomb, while other researchers hold the opinion that the graffiti was drawn later and not related to the construction and decoration of the tomb. However, the scholars agree that the name Kojimasis is Thracian [5–8]. The most interesting and best preserved is the frieze, with hunting scenes applied over several decorative bands in the upper part of the dome (Figure 3).

Figure 2. Part of the funeral feast scene from the chamber of the tomb at the village of Alexandrovo (author A. Cosentino, 2019).

Figure 3. Part of the frieze with hunting scenes from the chamber of the tomb at the village of Alexandrovo (author A. Cosentino, 2019).

Understandably, this frieze has aroused the greatest interest among scholars and the general public and gives the fullest picture of the skill of the artists who painted the tomb's interior in the late fourth or early third century BC.

In his publications, the discoverer of the tomb at the village of Alexandrovo, Georgi Kitov, mentions that during its exploration the fallen frescoes were carefully collected and sorted according to the place of discovery. Today the fragments of the frescoes are kept in the Museum Centre for Thracian Art in the Eastern Rhodopes near the village of Alexandrovo. Several of them have been analyzed within the current study.

While the tomb at the village of Alexandrovo is well studied from the historical and artistic point of view [5,9–15], analytical research on the painting materials that were used is limited to date. Some reports concerning the technological characteristics of monochrome decorative wall plasters are available in the literature [16–18]. Detailed knowledge of material composition is essential for art historical or archaeological research and for determining proper conservation and restoration procedures.

The aim of the present study is to identify, with the help of the combined analytical methods, the paint materials used in the decorative colored plasters in the interior of the Thracian tomb near the village of Alexandrovo, Bulgaria, as well as to clarify the technological features related to the making of the plaster. It should be noted that this is the first time that such a complex analytical approach has been applied to pigments and pigmented plasters from Thracian tombs. Thus, the current research will expand our knowledge of Thracian construction skills, as well as the ability of the Thracian elite to acquire, adapt, and further develop the tomb-building technologies known at the time.

2. Analytical Techniques Applied in the Study and Related Works

Since a single analytical technique cannot provide complete information about the materials and techniques used in the artwork, usually complementary methods of examination and analysis have to be carried out in parallel. Globally, in recent years, the multidisciplinary approach has been extensively used in the study of the composition of various materials of art and archaeological artifacts [19–21]. Because of the complex nature of the studied materials, in our study, we have adopted a multi-analytical approach to characterize the elemental composition, mineral phases, pigment, binder and filler materials, characteristic impurities and thermal properties of the plasters.

For the analysis of the elemental composition of micro samples, we have chosen Laser-Induced Breakdown Spectroscopy (LIBS). The capacity of LIBS for the identification of pigments in various painted artworks and cultural heritage objects has been demonstrated in many studies. In some studies, LIBS has been utilized for the characterization of pigments in the decorations and the colored layers of archaeological ceramics [22–25] and in the colored glaze of porcelain [26]. In the field of archaeometric analysis of pigments in murals and painted plasters, LIBS has been applied in several studies [27–31].

Paramagnetic electron resonance (EPR) spectroscopy finds application in archaeological science to study the origin of carbonate materials used in ancient marble sculptures [32,33] or as a preparation under layers in wooden picture masterpieces [34]. Investigation of different pigments in objects from cultural heritage could also be assisted by EPR spectroscopy. As many pigments originate from colored transition metal complexes or free radicals, the study of EPR of blue and green pigments [35,36] as well as of black pigments [32,37] can give additional knowledge of their composition or origin. In our study, EPR spectroscopy was used for the identification of the pigments and other materials in the composition of the preparatory layers and the decorative color plasters.

Infrared spectroscopy is a very useful analytical technique for both organic and inorganic materials in diverse archaeological and art contexts—building materials, mortars and mural decoration [38–42], icons, easel paintings and illuminated manuscripts [43–45], textile [46], archaeological and paleontological remains [47]. The attenuated total reflectance Fourier transform infrared (ATR-FTIR) technique offers the additional advantage of easy manipulation, chemically non-destructive analysis with a very small amount of sample or in particular cases—the possibility of non-invasive analysis. Moreover, combining spectral analysis with elemental composition data and dedicated spectral libraries allows the determination of a broad spectrum of materials. In the study of the mural decoration of Alexandrovo's tomb, the ATR-FTIR analysis provided simultaneous information on the presence of particular groups of inorganic materials such as silicate, carbonate, clay materials and organic matter in the studied samples.

The exact crystallographic and mineralogical composition was established based on the X-ray Powder Diffraction (XRD) method. X-ray powder diffraction analysis is widely applied in the diagnostics and conservation of cultural heritage materials to determine mineral compounds, textural properties of the crystalline phases, crystallite size and in-depth distribution [48,49]. Mural paintings from different periods were characterized successfully with XRD methods providing valuable information on mortar, pigment and filler composition [50–54].

The thermoanalytical methods, including thermogravimetry (TG), derivative thermogravimetry (DTG), differential thermal analysis (DTA), and differential scanning calorimetry (DSC), enable an analysis of painting media, wood and fossils in archaeological findings, manuscript parchments, mortars and pigments in historic buildings, synthetic polymer coatings, model and historic textiles and tapestries [55–57]. Although destructive in general, the thermal analysis requires only a few milligrams of sample and offers great sensitivity in terms of temperature and weight changes. Recently, it was successfully applied for the identification of organic additives in the mural decoration of Thracian monuments [17,18]. Because of the possibility of the method, in the current study, we have used the technique for the characterization of the thermal properties of the studied plasters and to support the identification of the pigments and the possible presence of an organic binder.

3. Materials and Methods

Small fragments (2–5 cm) representing all main colors encountered in the mural decoration of the tomb were collected and studied (Figure 4). Photos with respective scales are provided in the Supplementary Materials (Figures S1–S8). A series of powder samples from the colored plaster layers were obtained by scratching the painted surface of the fragments with a scalpel.

Figure 4. Representative color-plastered fragments from the Thracian tomb near the village of Alexandrovo, Bulgaria.

Laser-Induced Breakdown Spectroscopy (LIBS) and X-ray Powder Diffraction (XRD) analysis were carried out directly on the mural fragments. Paramagnetic Electron Resonance Spectroscopy (EPR), Attenuated Total Reflectance Fourier Transform Infrared Spectroscopy (ATR-FTIR) and thermal analysis measurements (DSC and TG analysis) were performed with the powder samples.

The LIBS analysis was performed by a Q-switched Nd:YAG laser. The laser was operated at a fundamental wavelength of 1064 nm with a pulse energy of 8 mJ and a pulse duration of 10 ns. The study was performed by applying 10 consecutive laser shots to the same location. For registration of the spectra, an Echelle spectrometer (Mechelle 5000, Andor Technology, Belfast, UK), with a grating of 52 grooves/mm and inverse linear dispersion of 1–4 nm/mm, was used. All spectra were registered at a delay time of 1 μs after the laser pulse and a gate of 1 μs. The resulting spectra were processed with Solis 64 software. The NIST atomic spectra database was used to identify the observed spectral lines and to determine the elements in the samples. A more detailed description of the experimental setup can be found in our previous work [58].

The EPR analysis was performed on a Bruker EPR spectrometer EMX Premium X working in the X-band at 9.4 GHz.

The materials underwent X-ray diffraction (XRD) examination using a Panalytical Empyrean diffractometer with Highscore Plus software (Scan Axis: Gonio; scan step length of 13.77 s with a continuous scan type; start position 5.0118 °2Th; end position 89.9868 °2Th;

step size 0.013 °2Th). For phase identification, a COD (Crystallography Open Database) database was used [59].

The FTIR and ATR-FTIR spectra were recorded on Bruker Invenio R FT spectrometer at a resolution of 2 cm^{-1} and 100 scans. The samples were measured in solid state (CsI pellet), and for the ATR technique, they were directly applied on a diamond crystal.

The TG-DSC analysis was carried out using a simultaneous thermal analyzer, the STA 449 F3 Jupiter (NETZSCH-Geratebau GmbH, Selb, Germany) apparatus, under the following conditions: temperature interval: 50 °C to 500 °C, temperature heating rate of 10 °C/min and static air atmosphere. Individual samples were placed in platinum pans.

4. Results

4.1. LIBS Analysis

LIBS analysis, in our case, is limited to the qualitative determination of elements in the investigated fragments due to the lack of suitable reference standards for plasters. However, due to the fact that the intensities of the spectral lines of a given element depend on its concentration in the sample, a semi-quantitative analysis can be made by determining the proportion of spectral line intensities for different samples. Although there are small differences in ablation in different fragments, the ratio of the intensities of the spectral lines is indicative of the relative abundance of the elements. In order to be able to compare the intensities of the spectral lines of the elements, the experimentally measured spectra of the samples were normalized by the total plasma emission intensity method.

LIBS analysis revealed the presence of Ca, Si, Mg, Al, Fe, Na, K, Mn and Sr in all of the samples but with different intensities of the spectral lines. A representative LIBS spectrum obtained for one of the analyzed fragments is demonstrated in Figure 5, and the spectral lines used to identify the elements are listed in Table S1.

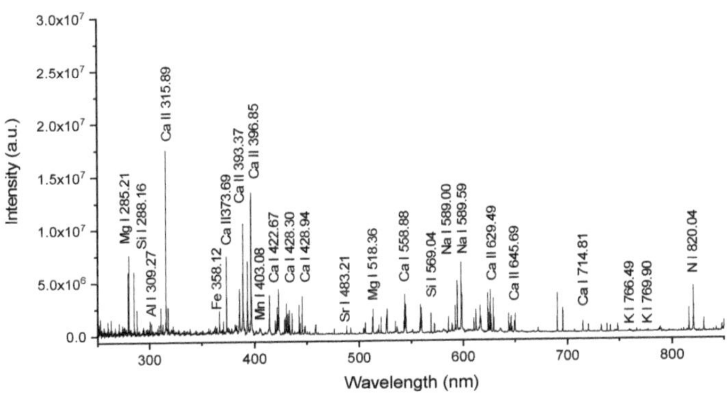

Figure 5. Exemplary LIBS spectrum obtained from one of the investigated fragments (red, dromos) representing the spectral range from 250 nm to 850 nm with some of the characteristic spectral lines of the detected elements.

It was noticed that in all of the fragments, the most abundant elements are Ca, Si and Mg. The presence of Ca and Si shows that the fragment material is a mixture mainly of calcium-bearing minerals and silicon-bearing minerals. This is a trivial finding because it is well-known that since ancient times mortar has been made commonly of lime (CaO), sand (SiO_2) and water [60]. The detection of strong magnesium spectral lines is not expected and assumes a significant presence of Mg-bearing minerals, most probably dolomite.

Iron is detected in all fragments (Figure 6), but the strongest iron spectral lines are observed in the red painted fragments (red, dromos and red, chamber). In the two black and in the brown painted fragments, the intensity of iron spectral lines decreases. The weakest iron spectral lines are detected in the spectra of white fragments (white, dromos and white, chamber) and the grey fragment (dark grey, dromos).

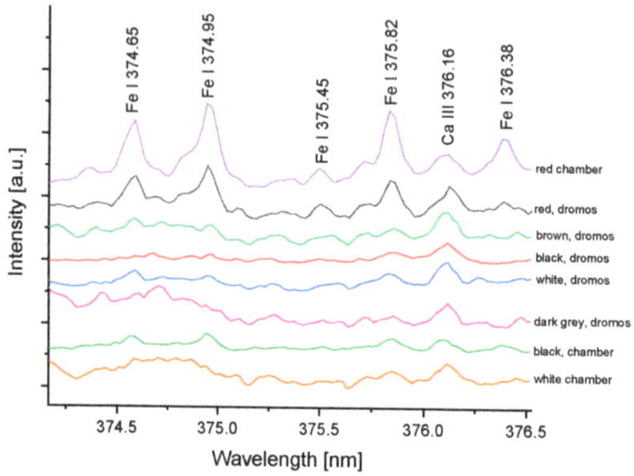

Figure 6. Section of LIBS spectrum from 374.5 nm to 376.5 nm showing an increased amount of Fe in colored layers of the fragments painted in red–red, chamber and red, dromos (the spectra are normalized to total plasma emission intensity).

To illustrate the above, a representative section of the spectra of two fragments with the most increased amount of iron oxides (red, chamber) and the one with the least (white, dromos) are compared in Figure S9a in the Supplementary Material. For comparison, iron and calcium spectral lines are designated. This reveals a considerable amount of iron oxides in the red-painted fragments, which most likely is red ochre considering that red ochre was commonly used in antiquity for red paint. It is interesting that in the red-painted fragments, an increased aluminum content compared to the other fragments was observed (Figure S9b), and this indicates the presence of aluminum-bearing minerals.

Kaolinite is known to be a commonly used material in ancient times as a white pigment, but it was also used as filler in paints [61]. Therefore, we can assume that kaolinite may have been added to the analyzed fragments to give a lighter shade to the red paint or was used as filler.

The black pigments used in antiquity are mainly carbon-based pigments in the form of charcoal, soot or graphite, manganese oxides and/or hydroxides and iron oxides [60–62]. In all of the analyzed fragments, manganese was observed only as a minor element. From this, it can be assumed that manganese is not intentionally added but rather naturally present in the raw material; hence, this element has no relation to the coloring of the plaster. Another possibility for the black coloration is carbon, but its detection is a challenging task due to its relatively poor spectrum and the low sensitivity of our analysis in the region of its analytical lines. Therefore, we cannot conclude with certainty whether carbon pigments or iron oxides were used for the black paint.

The detection of potassium and sodium in the spectra of all of the fragments could be attributed to the presence of alkali feldspars, which are one of the main components of mortars [63–65].

The presence of strontium in the mortars is not unusual since this element is naturally found in earth materials. Strontium often substitutes calcium in carbonates, sulfates, feldspars, mica, clays etc. [66–68].

4.2. EPR Analysis

The EPR analysis was carried out on a small amount of powder material, scraped from all types of layers building each one of the plasters. Figure 7, left, presents the general view of layers. It was established that most of the samples consist of three layers—coarse mortar, fine mortar and surface pigmented layer, while a "less colored underlayer" exists only in mortars with red color (red, dromos and red, chamber). Two samples deviate from the general appearance. The white sample from the dromos represents a one-layer slab with smoother upper and rougher lower surfaces. The brown sample from the dromos consists of only fine mortar with a pigmented upper part.

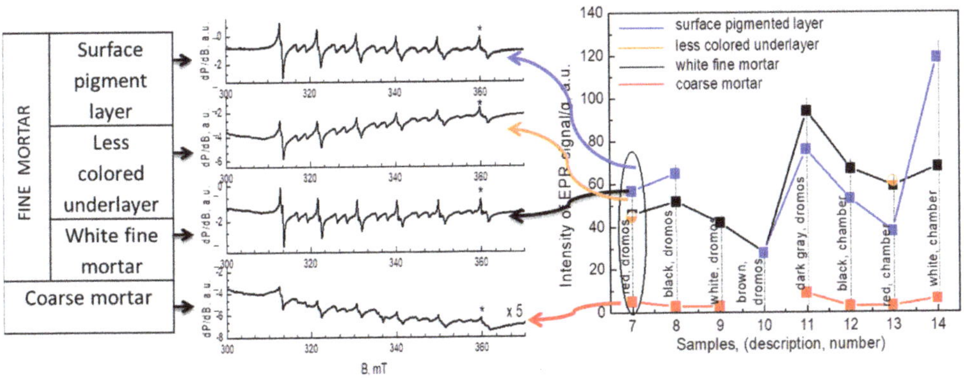

Figure 7. EPR analysis of surface-pigmented, fine and coarse mortar material sampled from fragments of the Alexandrovo tomb. Right—intensity of EPR signal of calcite in different layers, middle—EPR spectra of sample red, dromos; left—general view of the layers.

The EPR study has been focused on the investigation of carbonates, which are the main component of lime plasters. Although not paramagnetic, carbonates usually exhibit an EPR signal due to impurities of Mn^{2+} ions, which substitute for Ca^{2+} ions in the crystal lattice of calcite $CaCO_3$ and both for Ca^{2+} and Mg^{2+} ions in dolomite $CaMg(CO_3)_2$. Figure 8 presents the EPR spectra of Mn^{2+} ions in standard samples of calcite and dolomite that have been used as comparative spectra. The EPR analysis reveals that the portion taken from each layer is present only in the form of calcite or dominantly carbonate. To illustrate this assertion, Figure 6 presents spectra of isolated Mn^{2+} ions incorporated in the calcite used for the surface and fine mortar of the sample black, dromos.

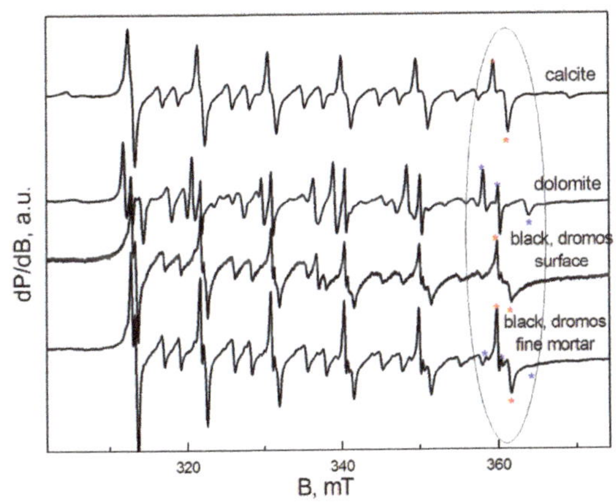

Figure 8. EPR spectra of calcite and dolomite standard minerals, the spectra of a pigmented surface and a fine mortar of sample black, dromos are also given.

The EPR spectrum of Mn^{2+} ions allows us not only to determine the chemical form of the carbonates but also to estimate the calcite content in each sample. Figure 7 shows the intensities of the sixth line in the spectrum of isolated ions (marked with *) recalculated for a 1 g sample. As seen, the amount of calcite is lower for the coarse mortar. The surface pigment layer and corresponding underlayer in red-colored samples (red, dromos and red, chamber) display identical calcite content. The last observation allows the conclusion that they were created by the same main material, and a pigment was added to the fine mortar in order to obtain a colored surface.

In the EPR spectra of the samples pigmented in black or brown color, one additional signal is observed between the third and fourth lines of the Mn^{2+} hyperfine structure. Figure 9 presents the spectra of samples black, dromos; brown, dromos; grey, dromos and black, chamber between 329 and 343 mT, as the extra signal not belonging to Mn^{2+} isolated ions is indicated. These signals exhibit g-factors varying from 2.0028 to 2.0031 and a narrow linewidth, ΔH_{pp}, of about 0.33–0.45 mT. The EPR parameters allow the signal to be related to the presence of radical species in the samples. Moreover, the established g-factor values are characteristic of carbon radicals [69], which was confirmed by the EPR analysis of a series of black pigments (Table S2). Previous studies of sources of black pigments, such as cremated bones [70], bituminous coal [71] and Beechwood [72], show very close values of g-factors (Table S2), and the only way to identify them is by using their linewidth values. The peak-to-peak linewidth in the spectra of radicals in Alexandrovo dark/brown colored fragments is in good agreement with the carbon radical signal width observed in materials of plant origin ashed at different temperatures, with the closest values found for beech wood [69]. The possibility of using, as a pigment, cremated animal bones or coal material is unlikely as their linewidth differs significantly from those we found in the studied plasters (Table S2).

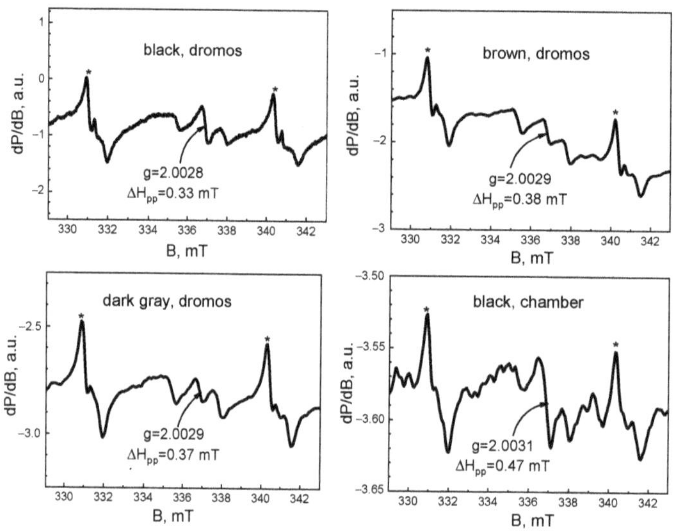

Figure 9. EPR spectra of black, brown and dark gray pigments from the Alexandrovo tomb.

4.3. *X-ray Diffraction Analysis*

The components in the pigmented layers and the underlayers lying beneath established by the XRD analysis are listed in Table 1. The XRD patterns are provided in the Supplementary Material (Figures S10–S17).

In both places—dromos and chamber, the painted layer is applied on top of a preparatory layer (fine mortar). The preparatory layer, lying beneath the paint, is mainly composed of lime mortar and, in the case of red layers, hematite and lime. In the lime used to prepare the plaster, there is a certain amount of aragonite, which is usually associated with the presence of fossils in the rock material from which the lime was prepared. In all cases where pigment was added, kaolinite is also present. As mentioned above, kaolinite was used as a white pigment and as a carrier of other pigments for better adhesion.

Table 1. Description of studied samples and components in the pigmented layers and the underlayers lying beneath established by XRD analysis.

Color	Location	Identified Components in the Pigments Layer	Identified Components in under Layer
Red	Dromos	Calcite ($CaCO_3$) Hematite (Fe_2O_3) Kaolinite ($Al_2Si_2O_5(OH)_4$)	Calcite ($CaCO_3$) Gypsum ($CaSO_4.2H_2O$) Kaolinite ($Al_2Si_2O_5(OH)_4$) Quartz (SiO_2) Hematite (Fe_2O_3)
Black	Dromos	Calcite ($CaCO_3$) Gypsum ($CaSO_4.2H_2O$) Muscovite ($KAl_2[AlSi_3O_{10}](OH)_2$)	Calcite ($CaCO_3$) Dolomite ($CaMg(CO_3)_2$) Aragonite ($CaCO_3$)
White	Dromos	Calcite ($CaCO_3$) Dolomite ($CaMg(CO_3)_2$)	
Grey	Dromos	Calcite ($CaCO_3$) Apatite ($Ca_5(PO_4)_3(F,Cl,OH)_2$)	Calcite ($CaCO_3$) Quartz (SiO_2) Bunsenite (NiO)
Brown	Dromos	Calcite ($CaCO_3$) Muscovite ($KAl_2[AlSi_3O_{10}](OH)_2$) Albite ($NaAlSi_3O_8$)	Calcite ($CaCO_3$)
Red	Chamber	Calcite ($CaCO_3$) Hematite (Fe_2O_3) Kaolinite ($Al_2Si_2O_5(OH)_4$) Aragonite ($CaCO_3$)	Calcite ($CaCO_3$) Aragonite ($CaCO_3$) Hematite (Fe_2O_3)
White	Chamber	Kaolinite ($Al_2Si_2O_5(OH)_4$) Gypsum ($CaSO_4.2H_2O$) Quartz (SiO_2)	
Black	Chamber	Graphite (C) Muscovite ($KAl_2[AlSi_3O_{10}](OH)_2$) Kaolinite ($Al_2Si_2O_5(OH)_4$) Gypsum ($CaSO_4.2H_2O$)	

The appearance of calcite in the diffraction patterns is explained by its presence in the plaster, which is a lime plaster and a thin layer on the surface of the frescoes formed over years as deposits from the groundwater. The presence of muscovite is attributed to soil contamination on the surface of the studied fragments. Quartz and albite are mineral components originating from the sand used in the coarse mortar.

The study of the phase composition of the pigments gives an idea of their mineral composition: white—kaolinite; black—graphite; red—hematite; dark grey—mixture of kaolinite with black pigment. In the latter case, the XRD analysis has indicated apatite presence, which is usually related to the use of bone black pigment. However, as the EPR analysis has shown that the carbon radicals in the dark grey fragment reveal characteristics closer to a carbon black pigment of plant origin, it should be considered that most probably the bone black pigment was added in smaller amounts or the apatite is present as a natural admixture of the white pigments. The following ICSD (Inorganic Crystal Structure Database) reference cards were used for phase identification: Calcite 98-015-8257; Aragonite 98-001-5194; Hematite 98-008-2135; Kaolinite 98-003-0285; Gypsum 98-002-7221; Quartz low 98-009-0145; Muscovite 98-002-6818; Albite 98-007-7421; Apatite 98-009-9362; Graphite 98-003-1829.

4.4. IR Analysis

The powder samples were examined by ATR Infrared spectroscopy to provide information on the presence of different inorganic and organic materials and, in this way, to identify the painting technique. The characteristic bands for the materials identified by the

ATR-FTIR analysis are summarized in Table S3. The analysis showed the content of a large amount of calcium carbonate in all tested samples. On the other hand, none of the spectra of the studied samples showed any organic matter.

The ATR-FTIR spectra of the powder samples collected from the white-colored plasters from dromos and chamber are shown in Figure 10. Both of these spectra evidenced the characteristic peaks of calcite at 1398–1407, 873, and 712 cm^{-1} attributed to the asymmetric stretching and out-of-plane and in-plane bending of CO_3^{2-} ions, respectively [73,74]. A weak band is observed at 2517 cm^{-1} also due to calcite [75]. The position of the bands at 872 and 712 cm^{-1} indicated that the samples contain mainly the calcite form of calcium carbonate and not the other polymorphic forms of aragonite or vaterite [76].

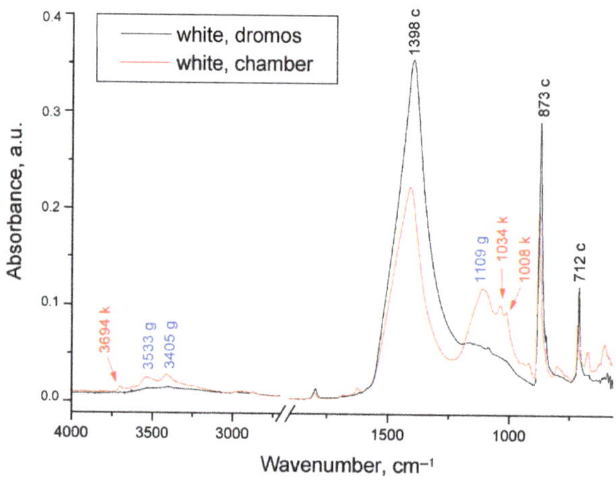

Figure 10. ATR-FTIR spectra of powder samples collected from the white-colored surface layers of dromos and chamber; the characteristic bands of identified materials are denoted by letters: c (calcite), k (kaolinite) and g (gypsum).

The analysis revealed that the powder sample collected from the white surface layer of the dromos solely contains calcite. However, the white-colored plaster from the chamber exhibits the presence of kaolinite in addition to calcite. The kaolinite was identified by its typical IR bands [77]: around 3600 cm^{-1} due to the OH stretchings, and a series of bands at 1034, 1008, and 797 cm^{-1} corresponding to Si-O stretching, Si-O-Al and Si-O-Si deformation vibrations of the aluminosilicates. In accordance with the XRD analysis, gypsum was identified in the white-colored plaster from the chamber based on the IR bands at 3533, 3405 and 1109 cm^{-1} [78,79].

A similar result was observed with the spectra of the two powder samples of black-colored plasters. The analysis of the spectra confirmed the presence of calcite as a major component in both samples, but the characteristic bands of kaolinite are observed in the spectrum of the black samples from the chamber while not present in the spectrum of the sample from the dromos (Figure S18).

Unlike the ATR spectra of the white and black colored plasters, the spectra of the samples from the red plasters (red, dromos and red, chamber) are almost completely identical (Figure S19). This fact emphasizes that the same red pigments were used in dromos and the chamber. In addition to the calcite, both spectra showed absorbance bands in positions that can be attributed to natural pigment red ochre [80,81].

Keeping in mind that the hematite is absorbing in the far-IR region, the spectra of the powder samples of red layers were measured in CsI pellet in an extended spectral range (4000–200 cm^{-1}) and compared to reference red ochre, hematite and calcite (Figure 11). A good correspondence was observed with the spectrum of reference red ochre (Kremer pigments, 40020) in the interval 1200–800 cm^{-1} and above 3200 cm^{-1}, although the spectrum was again dominated by the absorbance of the calcite. Hematite is characterized by strong IR bands at 532, 455 and 322 cm^{-1} [82]. The examined red layers showed IR bands at close positions, but this should be considered with caution as calcite and kaolinite also absorb at similar wavenumbers.

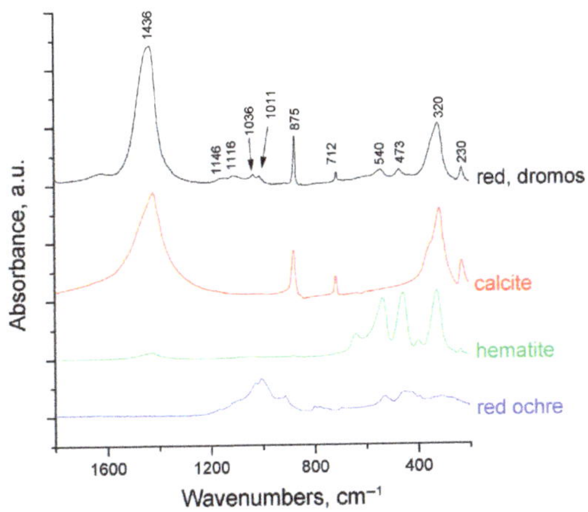

Figure 11. FTIR spectrum of powder sample collected from the red colored surface layer of domos, measured in CsI pellet, compared with reference red ochre, hematite and calcite.

The ATR-FTIR spectrum of the powder samples of the grey-colored layer from dromos (Figure S20) showed strong absorptions for calcite and, additionally, a band at 1028 cm^{-1}, which might be attributed to the stretching vibrations of apatite [83] if considered that IR bands around 3600 cm^{-1} for the OH stretching of silicates are missing in this case. In contrast, in the IR spectrum of the powder samples of the brown paint layer, bands around 3600 cm^{-1} and between 900 and 1220 cm^{-1} are visible as a result of silicate mineral presence such as kaolinite or muscovite [84].

4.5. DSC Analysis

Considering that the detection of organic matter could be significantly hampered by the strong absorptions of inorganic materials, such as calcite, in the IR spectra, further research into the possible presence of organic binder was conducted by TG-DSC analysis of some samples. The thermal analysis measurements (DSC and TG) of the powder samples from red paint layers are depicted in Figure 12.

The DSC curves showed only one well-defined endothermic peak with a maximum of about 116 °C and 119 °C for both studied samples. It could be assigned to evaporable water, absorbed in the clayey materials, and is in good agreement with the slight weight loss registered for these samples up to 150 °C (Figure 12). The position of the maxima may slightly shift depending on the proportions of the clayey minerals in the material composition [85]. Besides, the presence of this endothermic peak allows discrimination of the red ochre from other red pigments, such as red bolus and pure hematite, where no such low-temperature effect is observed [86]. Therefore, in the present case too, it could be concluded that the red pigment in the colored plaster is natural red ochre. At the same time, the slight weight loss registered above 150 °C (about 2%–2.5%) could be attributed to not well-defined endothermic processes due to the dehydration of both gypsum and/or kaolinite that are present in these samples [87,88].

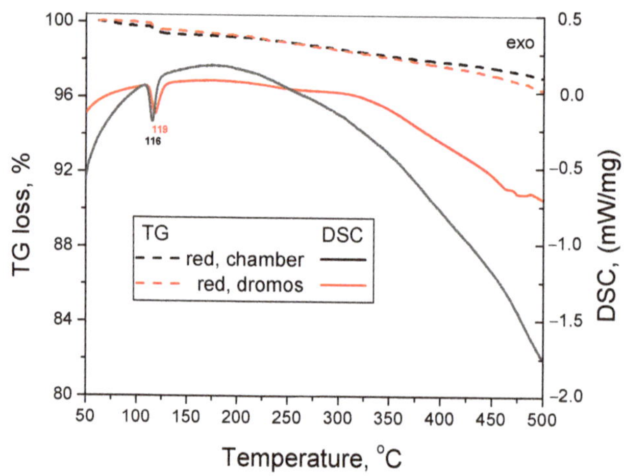

Figure 12. Thermal analysis measurements (DSC and TG) of the powder samples from red paint layers.

5. Discussion

According to historical evidence from ancient Greece and Rome, including the treatises of the Greek philosopher Theophrastus (371–287 BC), the Roman architect Vitruvius (80–15 BC), and the Roman scholar Pliny the Elder (23–79 AD) on plastering and coloring techniques, mortar was one of the main building materials used. It was applied either as a soldering substance between building elements (i.e., stone blocks, bricks) or as a plaster on the walls of the respective buildings. Different natural raw materials of local origin (such as limestone, dolomite, marble for the preparation of lime; sand from rock sources of different mineral composition, rock-forming, clay and micaceous minerals, etc.) were used to produce the mortar mixture, and the quality of the mortar was therefore strictly dependent on the ancient craftsmen's knowledge of the properties of minerals [89,90]. According to tradition, the decoration and painting of the walls was carried out by pre-treating them with successive application of several layers of mortar. Pigment was applied on top of the upper fine layer of mortar [91].

The results from the analytical study of the painting materials in the Thracian tomb near the village of Alexandrovo reveal the creative freedom of the ancient artist in applying wall decoration techniques. According to the analyzed data, in the case of the white (from the chamber), black, brown and grey decorations, the painting layer was applied on a two-layer lime plaster: one coarser layer between 0.5 and 0.8 cm thick and one finer layer between 0.15 and 0.30 cm thick. In the red decoration, three layers are clearly distinguishable: a rough layer (i.e., a lining plaster) about 0.8 cm thick; a white primer about 0.2 cm thick; and a light-colored fine layer less than 0.1 cm thick. The red wall decoration was carried out by applying pigments on top of a lightly colored pink primer.

The application of multiple layers of mortar is a crucial technological process that ensures even hardening of the plaster and determines its long-term stability. This process is a consequence of the properties of the slaked lime, as well as of the proportions between the composite materials in the mortar. The mineral composition of the fine mortar is a mixture of mainly quartz sand and calcite (as binder).

The painting layers' thickness, the presence of a large amount of calcium carbonate in all colored samples, and the absence of organic matter can be interpreted as evidence of the use of the fresco technique. This technique involves applying a water suspension of the pigment on wet, undried fine plaster. It was one of the most widely used in antiquity to paint large wall surfaces [92,93].

The pigment analysis showed that mostly natural pigments were used, including white—calcite and kaolinite; black pigment (amorphous C) of plant origin; red—natural ochres (colored earths), grey—amorphous C, possibly of mixed plant and bone material; and brown—mixture of brown-colored earths with amorphous C.

The identified components are summarized in Table 2 along with the analytical techniques supporting their presence.

Table 2. The identified components in studied samples from the dromos and chamber and the analytical techniques supporting their presence.

Color	Identified Components	LIBS	EPR	XRD	IR
Dromos					
Red	Calcite	+	+	+	+
	Hematite	+		+	+
	Kaolinite	+		+	+
	Quartz	+		+	
	Gypsum			+	+
Black	Calcite	+	+	+	+
	Gypsum			+	+
	Muscovite	+		+	
	Carbon		+		
	Dolomite	+	+	+	
White	Calcite	+	+	+	+
	Dolomite	+	+	+	
Grey	Calcite	+	+	+	+
	Apatite			+	+
	Carbon		+		
	Quartz	+		+	
Brown	Calcite	+	+	+	+
	Muscovite	+		+	
	Albite	+		+	
	Carbon		+		
Chamber					
Red	Calcite	+	+	+	+
	Hematite	+		+	+
	Kaolinite	+		+	+
	Aragonite	+		+	
White	Kaolinite	+		+	+
	Gypsum			+	+
	Quartz	+		+	
Black	Graphite		+	+	
	Muscovite	+		+	
	Kaolinite	+		+	+
	Gypsum	+		+	+

The abundance in nature, the wide variety of colors and shades, and excellent durability made the application of natural ochres as pigments in wall painting very common [80,81]. They are natural mineral mixtures containing silicates and clay minerals mixed with iron oxides—goethite, limonite and hematite. LIBS, XRD and IR analysis showed the red layers contain aluminosilicate material (kaolinite) and hematite as the materials producing the red color. The brown layers were found to contain aluminosilicate material, specifically muscovite, in addition to calcite, as indicated by LIBS and XRD analysis. The FTIR spectrum analysis also showed characteristic bands of aluminosilicate and calcite, with no contribution from any known brown pigments. The EPR analysis indicated the presence of carbon. Therefore, it is likely that the brown color in this case is produced by a mixture of brown-colored earths, such as muscovite with carbon.

6. Conclusions

For the first time, the mural paintings from one of the most significant Thracian tombs on Bulgarian territory situated near the village of Alexandrovo, Bulgaria, dating from the fourth century BC, were studied using complementary analytical methods. Through the application of Laser-Induced Breakdown Spectroscopy (LIBS), X-ray Powder Diffraction (XRD), Paramagnetic Electron Resonance Spectroscopy (ESR), Attenuated Total Reflectance Fourier Transform Infrared Spectroscopy (ATR-FTIR) and thermal analysis measurements (DSC and TG), a comprehensive understanding of the techniques and materials used was achieved.

The results demonstrated that the ancient Thracian artists were well-skilled in the creation of mural decoration based on the properties of lime, which they applied successively in several layers to achieve better carbonization, on the one hand, and uniform coloring with the pigment, on the other hand. The pigmented layer was applied on top of a fine mortar. The presence of calcite in all pigmented layers and the lack of organic binders indicated *fresco* as the main painting technique. The pigments used in the wall decoration are natural minerals suitable for the fresco technique because of their good chemical resistance. The color shades were achieved by using various pigment mixtures.

The current results contribute to the knowledge, research and study of monumental decoration and painting techniques acquired and developed in ancient times. They will allow the study of the similarities and distinctions of the Thracian tombs found on the territory of Bulgaria, as well as the Balkan region.

Supplementary Materials: The following supporting information can be downloaded at: https://www.mdpi.com/article/10.3390/min14040374/s1, Table S1. The spectral lines used for element detection in the LIBS analysis; Table S2. EPR parameters of reference black pigments and ashed materials; Table S3. Identified components and IR characteristic bands of the studied samples, measured by ATR-FTIR technique; Figures S1–S8 Photos of the studied fragment with respective scales; Figure S9. Two sections of LIBS spectra in the spectral region (a) 300 nm to 303 nm and (b) 394 nm to 395 nm obtained from fragments red, chamber; Figures S10–S17. Diffraction pattern of the studied samples; Figure S18. ATR-FTIR spectra of the black colored surface layers from dromos and chamber; and reference materials calcite, gypsum and kaolinite; Figure S19. ATR-FTIR spectra of the red-colored surface layers from dromos and chamber and reference calcite, gypsum and kaolinite; Figure S20. ATR-FTIR spectra of the dark grey colored surface layer from dromos and reference material hydroxyapatite.

Author Contributions: Conceptualization, G.N., R.S., B.S., G.A. and V.T.; formal analysis, G.N., R.S., B.S., G.A. and V.T.; funding acquisition, R.S. and G.N.; investigation, G.A., R.K., V.T., V.M., D.Y., M.D., G.N., R.S. and B.S.; methodology, G.A., V.M., G.N., R.S. and B.S.; project administration, G.N., R.S., B.S., G.A. and V.T.; supervision, G.N., R.S., B.S., G.A. and V.T.; validation, G.A., R.K., V.T., V.M., D.Y., M.D., G.N., R.S. and B.S.; visualization, G.A., R.K., V.T., V.M., D.Y., M.D., G.N., R.S. and B.S.; writing—original draft, G.A., R.K., V.T., V.M., D.Y., M.D., G.N., R.S. and B.S.; writing—review and editing, G.A., R.K., V.T., V.M., D.Y., M.D., G.N., R.S. and B.S. All authors have read and agreed to the published version of the manuscript.

Funding: This research was funded by Bulgarian Academy of Sciences under project "The Thracians—genesis and development of the ethnos, cultural identities".

Data Availability Statement: The data presented in this study are available upon request from the corresponding author.

Acknowledgments: Equipment of INFRAMAT (Research Infrastructure from National roadmap of Bulgaria), supported by Contract Д01-306/2021 with Bulgarian Ministry of Education and Science is used in the present investigations.

Conflicts of Interest: The authors declare no conflicts of interest.

References

1. Stoyanova, D. Tomb Architecture. In *A Companion to Ancient Thrace*; Valeva, J., Nankov, E., Graninger, D., Eds.; Wiley-Blackwell: Oxford, UK, 2015; pp. 158–179.
2. Kitov, G. A newly found Thracian tomb with frescoes. *Archael. Bulg.* **2001**, *5*, 15–29.

3. Kitov, G. *Alexandrovo Tomb (Original Title: Aleksandrovska Grobnitsa)*; Slavena: Varna, Bulgaria, 2005; ISBN 978-954-579-790-3.
4. Sultanov, A. The rock material used for the construction of the Thracian tomb near the village of Alexandrovo, Haskovsko (Original Title: Skalniat material, izpolzvan za stroezha na trakiyskata grobnitsa kray selo Aleksandrovo, Haskovo). *Stud. Mem. Rumen Katincharov Bull. Natl. Inst. Archaeol.* **2021**, *47*, 367–370.
5. Kitov, G. The Tomb in Alexandrovo (Original Title: Grobnitsata v Aleksandrovo). *Bull Mus. Hist.—Haskovo* **2004**, *2*, 149–175.
6. Gerassimova, V. Preliminary studies of the graffito from the Alexandrovo tomb. *Bull. Mus. Hist.—Haskovo* **2004**, *2*, 177–178.
7. Sharankov, N.A. Greek graffito in the Thracian tomb near Alexandrovo. *Archaeol. Bulg.* **2005**, *9*, 29–35.
8. Getov, L. On the graffiti from the Alexanderovo tomb (Original Title: Za grafita ot Aleksandrovskata grobnitsa). In *Phosphorion. Studia in Honorem Mariae Cicikova*; Gergova, D., Ed.; Prof. Marin Drinov: Sofia, Bulgaria, 2008; pp. 316–318.
9. Kitov, G. New observations in the Alexandrovo tomb. *Archeologia* **2004**, *45*, 42–51.
10. Kitov, G. New Discoveries in the Thracian tomb with frescoes by Alexandrovo. *Archaeol. Bulg.* **2005**, *9*, 15–28.
11. Kitov, G. Alexandrovo tomb—New facts and observations (Original Title: Aleksandrovskata grobnitsa—Novi fakti i nabliudenia). In *MIF 10: The Treasure*; Marazov, I., Ed.; New Bulgarian University: Sofia, Bulgaria, 2005; pp. 127–141.
12. Petrov, I. Thracian weaponry from the wall paintings of the tomb in the village of Alexandrovo (Original Title: Trakiyskoto vaorazhenie ot stenopisite na grobnitsata v Aleksandrovo). *Bull. Reg. Mus. Hist.—Haskovo* **2006**, *3*, 64–73.
13. Shalganova, T. Greek Hunting Myths and the Boar Hunt in the Alexandrpvo Tomb (Original Title: Gratskite Lovni Mitove i Lovat na Gligani ot Aleksandrovskata grobnitsa). In *MIF 9: Thrace and the Surrounding World*; Marazov, I., Ed.; New Bulgarian University: Sofia, Bulgaria, 2005; pp. 165–183.
14. Stoyanov, T. On the pictorial program of the Alexandrovo tomb. *Archeologia* **2008**, *49*, 58–67.
15. Nankov, E. Why one needs 'the odd man out'? The deer hunter with lagobolon from the frescoes in the Thracian tomb near Alexandrovo. *Archaeol. Bulg.* **2010**, *14*, 35–55.
16. Giurdzhiiska, D. Technological observations on the painting techniques at Alexandrovo tomb. The problem of the organic binding medium in the Antiquity. *Bulg. E-J. Archeol.* **2013**, *3*, 137–154.
17. Guirdzhiiska, D.; Zlateva, B.; Glavcheva, Z. Polished Decorative Fields in Thracian Fresco Tombs from the Hellenistic period—Archaeometrical Research. *STAR Sci. Techn. Archaeol. Res.* **2017**, *3*, 428–436. [CrossRef]
18. Glavcheva, Z.; Yancheva, D.; Velcheva, E.; Stamboliyska, B.; Petrova, N.; Lalev, G.; Todorov, V. Analytical studies of the Alexandrovo Thracian tomb wall paintings. *Spectrochim. Acta A* **2016**, *152*, 622–628. [CrossRef] [PubMed]
19. Mazzeo, R. (Ed.) Analytical Chemistry for Cultural Heritage. In *Topics in Current Chemistry Collections*; Springer: Berlin/Heidelberg, Germany, 2017.
20. Madariaga, J.M. Analytical chemistry in the field of cultural heritage. *Anal. Methods* **2015**, *7*, 4848–4876. [CrossRef]
21. Siddall, R. Mineral pigments in archaeology: Their analysis and the range of available materials. *Minerals* **2018**, *8*, 201. [CrossRef]
22. . López, A.J.; Nicolás, G.; Mateo, M.P.; Piñón, V.; Tobar, M.J.; Ramil, A. Compositional analysis of Hispanic Terra Sigillata by laser-induced breakdown spectroscopy. *Spectrochim. Acta B* **2005**, *60*, 1149–1154. [CrossRef]
23. Oztoprak, B.G.; Sinmaz, M.A.; Tülek, F. Composition analysis of medieval ceramics by laser-induced breakdown spectroscopy (LIBS). *Appl. Phys. A* **2016**, *122*, 557. [CrossRef]
24. Osticioli, I.; Agresti, J.; Fornacelli, C.; Memmi, I.T.; Siano, S. Potential role of LIPS elemental depth profiling in authentication studies of unglazed earthenware artifacts. *J. Anal. At. Spectrom.* **2012**, *27*, 827–833. [CrossRef]
25. Madkour, F.; Imam, H.; Elsayed, K.; Meheina, G. Elemental Analysis Study of Glazes and Ceramic Bodies from Mamluk and Ottoman Periods in Egypt by Laser-Induced Breakdown Spectroscopy (LIBS). *Period. Mineral.* **2015**, *84*, 107–121.
26. Sadek, H.; Simileanu, M.; Radvan, R.; Goumaa, R. Identification of porcelain pigments by Laser Induced Breakdown Spectroscopy. *J. Optoelectron. Adv. Mater.* **2012**, *4*, 858–862.
27. Sun, D.; Zhang, Y.; Yin, Y.; Zhang, Z.; Qian, H.; Wang, Y.; Yu, Z.; Su, B.; Dong, C.; Su, M. A Comparative Study of the Method to Rapid Identification of the Mural Pigments by Combining LIBS-Based Dataset and Machine Learning Methods. *Chemosensors* **2022**, *10*, 389. [CrossRef]
28. Brysbaert, A.; Melessanaki, K.; Anglos, D. Pigment analysis in Bronze Age Aegean and Eastern Mediterranean painted plaster by laser-induced breakdown spectroscopy (LIBS). *J. Archaeol. Sci.* **2006**, *33*, 1095–1104. [CrossRef]
29. Bruder, R.; Detalle, V.; Coupry, C. An example of the complementarity of laser- induced breakdown spectroscopy and Raman microscopy for wall painting pigments analysis. *J. Raman Spectrosc.* **2007**, *38*, 909–915. [CrossRef]
30. Westlake, P.; Siozos, P.; Philippidis, A.; Apostolaki, C.; Derham, B.; Terlixi, A.; Perdikatsis, V.; Jones, R.; Anglos, D. Studying pigments on painted plaster in Minoan, Roman and early Byzantine Crete. A multi-analytical technique approach. *Anal. Bioanal. Chem.* **2012**, *402*, 1413–1432. [CrossRef] [PubMed]
31. Toschi, F.; Paladini, A.; Colosi, F.; Cafarelli, P.; Valentini, V.; Falconieri, M.; Gagliardi, S.; Santoro, P. A multi-technique approach for the characterization of Roman mural paintings. *Appl. Surf. Sci.* **2013**, *284*, 291–296. [CrossRef]
32. Weihe, H.; Piligkos, S.; Barra, A.L.; Laursen, I.; Johnse, O. EPR of Mn^{2+} impurities in calcite: A detailed study pertinent to marble provenance determination. *Archaeometry* **2009**, *51*, 43–48. [CrossRef]
33. Attanasio, D.; Platania, R. ESR Spectroscopy as a Tool for Identifying Joining Fragments of Antique Marbles: The Example of a Pulpit by Donatello and Michelozzo. *J. Magn. Res.* **2000**, *144*, 322–329. [CrossRef] [PubMed]

34. Romanelli, M.; Buccianti, A.; Benedetto, F.; Bellucci, L.; Cemicky, S. An innovative electron paramagnetic resonance and statistical analysis approach to investigate the geographical origin of multi-layered samples from a Renaissance painting. *Microchem. J.* **2022**, *177*, 107219. [CrossRef]
35. Kuzio, O.R.; Hornak, J.P. An Introduction to the Electron Paramagnetic Resonance Spectral Library of Pigments. *Heritage* **2022**, *5*, 545–566. [CrossRef]
36. Javier, S.; Horna, J.P. A nondestructive method of identifying pigments on canvas using low frequency electron paramagnetic resonance spectroscopy. *J. Amer. Inst. Conserv.* **2018**, *57*, 73–82. [CrossRef]
37. Le Pape, L. Application of EPR in Studies of Archaeological Samples. In *Modern Magnetic Resonance*; Webb, G., Ed.; Springer: Berlin/Heidelberg, Germany, 2016.
38. Bitossi, G.; Giorgi, R.; Mauro, M.; Salvadori, B.; Dei, L. Spectroscopic Techniques in Cultural Heritage Conservation: A Survey. *Appl. Spectrosc. Rev.* **2005**, *40*, 187–228. [CrossRef]
39. Cirlini, M.; Graiff, C.; Lottici, P.P.; Palla, G.; Casoli, A.; Bergamonti, L. Characterization of Waxes in the Roman Wall Paintings of the Herculaneum Site (Italy). *Appl. Sci.* **2022**, *12*, 11264. [CrossRef]
40. Brecoulaki, H.; Zaitoun, C.; Stocker, S.R.; Davis, J.L.; Karydas, G.; Colombini, M.P.; Bartolucci, U. An Archer from the Palace of Nestor: A New Wall-Painting Fragment in the Chora Museum. *Hesperia* **2008**, *77*, 363–397. [CrossRef]
41. Bakiler, M.; Kırmızı, B.; Öztürk, Ö.O.; Hanyalı, Ö.B.; Çağlar, E.; Köroğlu, G. Material characterization of the Late Roman wall painting samples from Sinop Balatlar Church Complex in the black sea region of Turkey. *Microchem. J.* **2016**, *126*, 263–273. [CrossRef]
42. Daniilia, S.; Minopoulou, E.; Andrikopoulos, K.S.; Tsakaof, A.; Bairachtari, K. From Byzantine to post-Byzantine art: The painting technique of St Stephen's wall paintings at Meteora. *J. Amer. Sci.* **2008**, *35*, 2474–2485. [CrossRef]
43. Derrick, M.R.; Stulik, D.C.; Landry, J.M. Infrared Spectroscopy in Conservation Science. In *Scientific Tools for Conservation*; Getty Conservation Institute: Los Angeles, CA, USA, 1999.
44. Daher, C.; Paris, C.; Le Ho, A.-S.; Bellot-Gurlet, L.; Echard, J.-P. A Joint use of Raman and Infrared Spectroscopies for the Identication of Natural Organic Media Used in Ancient Varnishes. *J. Ram. Spectr.* **2010**, *41*, 1204–1209. [CrossRef]
45. Bacci, M.; Picollo, M.M. Non-Destructive Spectroscopic Detection of Cobalt (II) in Paintings and Glass. *Stud. Conserv.* **1996**, *121*, 136–144. [CrossRef]
46. Peets, P.; Kaupmees, K.; Vahur, S.; Leito, I. Refectance FT-IR spectroscopy as a viable option for textile fiber identification. *Herit. Sci.* **2019**, *7*, 93–103. [CrossRef]
47. Koupadi, K.; Boyatzis, S.C.; Roumpou, M.; Kalogeropoulos, N. Organic Remains in Early Christian Egyptian Metal Vessels: Investigation with Fourier Transform Infrared Spectroscopy and Gas Chromatography–Mass Spectrometry, Despoina Kotzamani. *Heritage* **2021**, *4*, 3611–3629. [CrossRef]
48. Artioli, G. Science for the cultural heritage: The contribution of X-ray diffraction. *Rend. Fis. Acc. Lincei* **2013**, *24* (Suppl. S1), 55–62. [CrossRef]
49. Gonzalez, V.; Cotte, M.; Vanmeert, F.; de Nolf, W.; Janssens, K. X-ray Diffraction Mapping for Cultural Heritage Science: A Review of Experimental Configurations and Applications. *Chem. Eur. J.* **2019**, *25*, 1–19. [CrossRef]
50. Pagès-Camagna, S.; Laval, E.; Vigears, D. Non-destructive and in situ analysis of Egyptian wall paintings by X-ray diffraction and X-ray fluorescence portable systems. *Appl. Phys. A* **2010**, *100*, 671–681. [CrossRef]
51. Duran, A.; Castaing, J.; Walter, P. X-ray diffraction studies of Pompeian wall paintings using synchrotron radiation and dedicated laboratory made systems. *Appl. Phys. A* **2010**, *99*, 333–340. [CrossRef]
52. Duran, A.; Perez-Rodriguez, J.L.; Jimenez de Haro, M.C.; Franquelo, M.L.; Robador, M.D. Analytical study of Roman and Arabic wall paintings in the Patio De Banderas of Reales Alcazares' Palace using non-destructive XRD/XRF and complementary techniques. *J. Archaeolog. Sci.* **2011**, *38*, 2366–2377. [CrossRef]
53. Iordanidis, A.; Garcia-Guinea, J.; Strati, A.; Gkimourtzina, A.; Papoulidou, A. Byzantine wall paintings from Kastoria, northern Greece: Spectroscopic study of pigments and efflorescing salts. *Spectrochim. Acta Part A* **2011**, *78*, 874–887. [CrossRef]
54. Mohanu, I.; Mohanu, D.I.; Gomoiu, I.; Barbu, O.-H.; Fechet, R.-M.; Vlad, N.; Voicu, G.; Truşcă, R. Study of the frescoes in Ioneştii Govorii wooden church (Romania) using multi-technique investigations. *Microchem. J.* **2016**, *126*, 332–340. [CrossRef]
55. Pires, J.; Cruz, A.J. Techniques of thermal analysis applied to the study of cultural heritage. *J. Therm. Anal. Calorim.* **2007**, *87*, 411–415. [CrossRef]
56. Fessas, D.; Schiraldi, A.; Tenni, R.; Zuccarello, L.V.; Bairati, A.; Facchini, A. Calorimetric, biochemical and morphological investigations to validate a restoration method of fire injured ancient parchment. *Thermochim. Acta* **2000**, *348*, 129–137. [CrossRef]
57. Tomassetti, M.; Marini, F.; Campanella, L.; Coppa, A. Archaeometric classification of ancient human fossil bones, with particular attention to their carbonate content, using chemometrics, thermogravimetry and ICP emission. *Chem. Centr. J.* **2014**, *8*, 26. [CrossRef] [PubMed]
58. Tankova, V.; Mihailov, V.; Malcheva, G.; Penkova, P.; Leshtakov, L. Quantitative determination of antimony in archaeological bronze artefacts by laser-induced breakdown spectroscopy. *J. Phys. Conf. Ser.* **2021**, *1859*, 012026. [CrossRef]
59. Grazulis, S.; Chateigner, D.; Downs, R.T.; Yokochi, A.T.; Quiros, M.; Lutterotti, L.; Manakova, E.; Butkus, J.; Moeck, P.; Le Bail, A. Crystallography Open Database—An open-access collection of crystal structures. *J. Appl. Cryst.* **2009**, *42*, 726–729. [CrossRef]
60. Goffer, Z. *Archaeological Chemistry*, 2nd ed.; John Wiley & Sons, Inc.: Hoboken, NJ, USA, 2007.

61. Hradil, D.; Hradilová, J.; Bezdička, P. Clay Minerals in European Painting of the Mediaeval and Baroque Periods. *Minerals* **2020**, *10*, 255. [CrossRef]
62. Domingo, I.; Chieli, A. Characterizing the pigments and paints of prehistoric artists. *Archeol. Anthrop. Sci.* **2021**, *13*, 196. [CrossRef]
63. Borsoi, G.; Santos Silva, A.; Menezes, P.; Candeias, A.; Mirão, J. Analytical characterization of ancient mortars from the archaeological roman site of Pisões (Beja, Portugal). *Constr. Build. Mater.* **2019**, *204*, 597–608. [CrossRef]
64. Gheris, A. New dating approach based on the petrographical, mineralogical and chemical characterization of ancient lime mortar: Case study of the archaeological site of Hippo, Annaba city, Algeria. *Herit. Sci.* **2023**, *11*, 103. [CrossRef]
65. Fort, R.; Varas-Muriel, M.J.; Ergenç, D.; Cassar, J.; Anastasi, M.; Vella, N.C. The technology of ancient lime mortars from the Żejtun Roman Villa (Malta). *Archeol. Anthrop. Sci.* **2023**, *15*, 15. [CrossRef]
66. Lindroos, A.-J.; Aro, L. Natural factors influencing strontium concentrations in bulk and throughfall deposition, soil solution and litterfall in forest ecosystems on Olkiluoto Island, southwestern Finland. *Boreal Environ. Res.* **2019**, *24*, 233–242.
67. Anderson, E.; Almond, M.J.; Matthews, W. Analysis of wall plasters and natural sediments from the Neolithic town of Çatalhöyük (Turkey) by a range of analytical techniques. *Spectrochim. Acta A* **2014**, *133*, 326–334. [CrossRef] [PubMed]
68. Lerouge, C.; Gaucher, E.C.; Tournassat, C.; Negrel, P.; Crouzet, C.; Guerrot, C.; Gautier, A.; Michel, P.; Vinsot, A.; Buschaert, S. Strontium distribution and origins in a natural clayey formation (Callovian-Oxfordian, Paris Basin, France): A new sequential extraction procedure. *Geochim. Cosmochim. Acta* **2010**, *74*, 2926–2942. [CrossRef]
69. Tian, L.; Koshland, C.P.; Yano, J.; Yachandra, V.K.; Yu, I.T.; Lee, S.C.; Lucas, D. Carbon-centered free radicals in particulate matter emissions from wood and coal combustion. *Energy Fuels* **2009**, *23*, 2523–2526. [CrossRef]
70. Iriartea, E.; García-Tojalb, J.; Santanac, J.; Jorge-Villard, S.E.; Teiraf, L.; JMuñizg, J.; Ibañezh, J. Geochemical and spectroscopic approach to the characterization of earliest cremated human bones from the Levant (PPNB of Kharaysin, Jordan). *J. Archaeol. Sci. Rep.* **2020**, *30*, 102211. [CrossRef]
71. Retcofsky, H.L.; Stark, J.M.; Friedel, R.A. *Electron Spin Resonance in American Coals*; U.S. Department of the Interior, Bureau of Mines, Pittsburgh Coal Research Center: Pittsburgh, PA, USA, 1968.
72. Trubetskaya, A.; Jensen, P.A.; Jensen, A.D.; Glarborg, P.; Larsen, F.H. Characterization of free radicals by electron spin resonance spectroscopy in biochars from pyrolysis at high heating rates and at high temperatures Mogens Larsen Andersen. *Biomass Bioenergy* **2016**, *94*, 117–129. [CrossRef]
73. Gunasekaran, S.; Anbalagan, G. Spectroscopic characterization of natural Calcite Minerals. *Spectrochim. Acta A* **2007**, *68*, 656–664. [CrossRef] [PubMed]
74. Gunasekaran, S.; Anbalagan, G.; Pandi, S. Raman and infrared spectra of carbonates of calcite structure. *J. Raman Spectrosc.* **2006**, *37*, 892–899. [CrossRef]
75. Lan, J.; Lv, J.; Zhang, W.; Feng, J.; Liu, Y.; Wang, Z.; Zhao, M. Discrimination of Types of Thermal Sensitive Paper by Characterization of the Inorganic Additives by Fourier Transform Infrared Microscopy, Confocal Raman Microscopy, and Scanning Electronic Microscope-Energy Dispersive Spectroscop. *Analyt. Lett.* **2012**, *45*, 2763–2773. [CrossRef]
76. Vagenas, N.V.; Gatsouli, A.; Kontoyannis, C.G. Quantitative analysis of synthetic calcium DSC carbonate polymorphs using FT-IR spectroscopy. *Talanta* **2003**, *59*, 831–836. [CrossRef]
77. Moenke, H. Silica, the three-dimensional silicates, borosilicates and beryllium silicates. In *The Infrared Spectra of Minerals*; Farmer, V.C., Ed.; Mineralogical Society: London, UK, 1974; pp. 365–382.
78. Anbalagan, G.; Mukundakumari, S.; Murugesan, K.S.; Gunasekaran, S. Infrared, optical absorption, and EPR spectroscopic studies on natural gypsum. *Vibr. Spectrosc.* **2009**, *50*, 226–230. [CrossRef]
79. Vahur, S.; Teearu, A.; Peets, P.; Joosu, L.; Leito, I. ATR-FT-IR spectral collection of conservation materials in the extended region of 4000-80 cm^{-1}. *Anal Bioanal Chem.* **2016**, *408*, 3373–3379. [CrossRef]
80. Hradil, D.; Grygar, T.; Hradilova, J.; Bezdicka, P. Clay and iron oxide pigments in the history of painting. *Appl. Clay Sci.* **2003**, *22*, 223–236. [CrossRef]
81. Perdikatsis, V.; Brecoulaki, H. The Use of Red and Yellow Ochres as Painting Materials in Ancient Macedonia. In Proceedings of the 4th Symposium of the Hellenic Society for Archaeometry, Athens, Greece, 28–31 May 2003; National Hellenic Research Foundation, Facorellis, Y., Zacharias, N., Polikreti, K., Eds.; BAR International Series 1746; Archaeopress Publishers of British Archaeological Reports: Oxford, UK; pp. 559–569.
82. Bikiaris, D.; Daniilia, S.; Sotiropoulou, S.; Katsimbiri, O.; Pavlidou, E.; Moutsatsou, A.P.; Chryssoulakis, Y. Ochre differentiation through micro-Raman and micro-FTIR spectroscopies. *Spectrochim. Acta Part A* **1999**, *56*, 3–18. [CrossRef]
83. Hossain, M.S.; Ahmed, S. FTIR spectrum analysis to predict the crystalline and amorphous phases of hydroxyapatite: A comparison of vibrational motion to reflection. *RSC Adv.* **2023**, *13*, 14625–14630. [CrossRef]
84. Sontevska, V.; Gligor Jovanovski, G.; Makreski, P.; Raskovska, A.; Soptrajanov, B. Minerals From Macedonia. XXI. Vibrational Spectroscopy as Identificational Tool for Some Phyllosilicate Minerals. *Acta Chim. Slov.* **2008**, *55*, 757–766.
85. Ndjigui, P.; Mbey, J.A.; Nzeukou, A.N. Mineralogical, physical and mechanical features of ceramic products of the alluvial clastic clays from the Ngog-Lituba region, Southern Cameroon. *J. Build. Eng.* **2016**, *5*, 151–157. [CrossRef]
86. Čiuladienė, A.; Luckutė, A.; Kiuberis, J.; Kareiva, A. Investigation of the chemical composition of red pigments and binding media. *CHEMIJA* **2018**, *29*, 243–256. [CrossRef]
87. Engbrecht, D.C.; Hirschfeld, D.A. Thermal analysis of calcium sulfate dihydrate sources used to manufacture gypsum wallboard. *Thermochim. Acta* **2016**, *639*, 173–185. [CrossRef]

88. Dohnalová, Ž.; Svoboda, L.; Šulcová, P. Characterization of kaolin dispersion using acoustic and electroacoustic spectroscopy. *J. Min. Metall.* **2008**, *44*, 63–72. [CrossRef]
89. Hodges, H. *Artifacts: An Introduction to Early Materials and Technology*; Duckworth & Co Ltd.: London, UK, 1989.
90. Boness, D.; Panagiotopoulos, D.; Goren, Y. Minoan plaster technology as evident from the 'precinct' structure at Koumasa, Crete: A microarchaeological study. *J. Archaeol. Sci. Rep.* **2017**, *14*, 392–408. [CrossRef]
91. Piovesan, R.; Siddall, R.; Mazzoli, C.; Nodari, L. The Temple of Venus (Pompeii): A study of the pigments and painting technique. *J. Archaeol. Sci.* **2011**, *38*, 2633–2643. [CrossRef]
92. Piovesan, R.; Mazzoli, C.; Maritan, L.; Cornale, P. Fresco and lime-paint: An experimental study and objective criteria for distinguishing between these painting techniques. *Archaeometry* **2012**, *54*, 723–736. [CrossRef]
93. Regazzoni, L.; Cavallo, G.; Biondelli, D.; Gilardi, J. Microscopic Analysis of Wall Painting Techniques: Laboratory Replicas and Romanesque Case Studies in Southern Switzerland. *Stud. Conserv.* **2018**, *63*, 326–341. [CrossRef]

minerals

MDPI

Article

3D Modelling Approach to Enhance the Characterization of a Bronze Age Nuragic Site

Stefano Cara [1], Paolo Valera [2] and Carlo Matzuzzi [1,*]

1 National Research Council of Italy—Institute of Environmental Geology and Geoengineering (IGAG), 09124 Cagliari, Italy; stefano.cara@cnr.it
2 Department of Civil and Environmental Engineering and Architecture (DICAAR), University of Cagliari, 09124 Cagliari, Italy; pvalera@unica.it
* Correspondence: carlo.matzuzzi@igag.cnr.it

Abstract: Megalithism in Sardinia (Italy) had its highest expression during the Bronze Age with the creation of monumental complexes known as Nuraghes. These unique monuments have recently been the subject of in-depth investigations for their potential to be recognized as World Heritage Sites (by UNESCO). The main purpose of our research was to make a contribution to obtain a more in-depth characterization of these monuments by testing a 3D model of a complex Nuraghe, integrated with an analysis of the geolithological context. This work first focused on the geological and typological investigation of the materials used in its construction, which was then compared with the geolithological characteristics of the region. A survey of the outcropping remains was carried out by means of Structure-from-Motion Multi-View Stereo (SfM-MVS) photogrammetry with UAV ground and aerial acquisition using APS-C photo sensors, georeferenced with an RTK-GNSS ground survey. The level of accuracy of our digital models shows the potential of the proposed method, giving accurate and geometrically consistent 3D reconstructions in terms of georeferencing error, shape and surface. The survey method allows for the virtualization of the current state of conservation of the Nuraghe, giving a solid basis to set up further (future) archaeological excavations and to contribute to knowledge on the architecture of the structures. This study also provides useful information on the nature and origin of the construction materials and proposes a hypothesis on the original dimensions of the monument, which is often a topic of debate in the world of archaeology.

Keywords: UAV; RPAS; photogrammetry; geoarchaeology; 3D reconstruction; virtual archaeology; Nuraghe

1. Introduction

The purpose of this work was to obtain a more in-depth characterization of Nuraghes, the unique and identifying monuments of the Nuragic civilization developed during the Bronze Age in Sardinia (Italy). These monuments are currently the subject of in-depth scientific studies for the purpose of their recognition as UNESCO World Heritage Sites. This study's results allowed for the development of an experimental survey method based on UAV platform-based photogrammetric survey and 3D modelling techniques applied to a complex Nuraghe. This study was integrated with an analysis of the geolithological context to provide useful information on the nature and origin of the building materials and to propose a realistic hypothesis on the original proportions and dimensions of the monument, based on the morphometric analysis of the virtual models obtained. The Nuragic civilization developed in Sardinia mainly during the Bronze Age (18th–10th century BC). The naming of this civilization does not refer directly to an ethnic subject nor to a major ideal current but rather to the impressive and unique architectural phenomenon of megalithic towers, known in the local language as "Nuraghes" [1–4].

The Nuraghes, with about 8000 currently characterized in Sardinia, have undergone a remarkable architectural evolution, starting from the earliest and simplest form of so-called "archaic Nuraghes". It is precisely from the evolution of these archaic Nuraghes—in a phase of the Middle Bronze Age about which scholars still debate, between the 17th and 14th centuries BC—that led to the definition of the tower-shaped nuraghe form, with a large central chamber [5]. This first phase of evolution was followed by a second one (late 14th–11th century BC), in which the building knowledge of the Nuragic people became intertwined with the experiences of Aegean architects, since circular chambers, corridors, and large ogival vaults are common to Sardinia, the island of Crete, and mainland Greece [6,7]. Nuraghes can be mainly represented by three types: the so-called archaic Nuraghes, developed from the beginning on one level with corridors and chambers without a vaulted enclosure; the single-tower Nuraghe, containing one or more vaulted chambers; and the more evolved construction of the complex polylobate Nuraghe [5]. Thus, the basic construction concept of the most evolved Nuraghe is centered on the construction of the truncated conical tower with a chamber, made using the technique of drystone masonry by laying concentric rows of polygonal stone boulders, slightly projecting towards the center, until obtaining the vaulted closure of the chamber, called "a tholos". The use of the term "tholos" in our paper refers to the construction characteristic of the architectural structure. This term is used extensively in archaeology, and particularly by scholars of Sardinian archaeology, in reference to the construction of Nuragic tower chambers to refer to a circular, truncated-cone-shaped construction consisting of rings of projecting stone blocks forming a pseudo-dome [8]. The megalithic structure of the wall usually consists of blocks with decreasing size from the bottom to the top: the larger ones are arranged according to a so-called "cyclopean" polygonal structure and are arranged in more or less regular overlapping rows to form the base, while for the upper part of the masonry, elements in many cases are placed in an "isodomic" arrangement, so as to obtain a better distribution of loads and more easily allow for the curved profile of the vault [5]. For the construction of Nuraghes, local stone materials were usually used. The nature and characteristics of these materials certainly influenced the aesthetic appearance and durability of the constructions, but they certainly did not prevent ancient builders from constructing their monuments even with materials that were difficult to work with, while maintaining the basic architectural and structural characteristics. The construction of the walls is in itself very complex and involved not only the laying of boulders of considerable size but also the use of stones mixed with earth of various nature and granulometry, intended to fill in the voids, to make the wall structure more homogeneous and stable and insulate it from the outside. The outer walls of the towers show a varied inclination (usually less than 10°), which in some cases is not constant but decreases strongly in the upper part of the structure (as found in the Nuraghe Nuraddeo of Suni [5]). The module of the inner tholos chamber was generally replicated upward with the same proportions, but with reduced dimensions, and could also develop on three superimposed levels (such as the Nuraghe Santu Antine in Torralba) giving rise to towers of considerable height. However, to date, in no Nuraghe has the summit part (probably consisting of a terrace) been found intact; therefore, it is not possible to define exactly the maximum height originally reached by the Nuragic towers. Hypotheses about the characteristics of the summit structure have been made by past studies based on stylized Nuraghe artefacts (bronze and lithic) that show the presence above the towers of a terrace with a balcony projecting from the masonry and supported by brackets, found, for example, in large numbers in the archaeological area of "Su Nuraxi" in Barumini [1,9].

The typical structure of the Nuragic tower includes at least three basic characterizing elements: the access door lintel, the access corridor, and the so called "a tholos" chamber. The other distinctive element is the staircase, usually with helicoidal development and built within the masonry of the tower itself, allowing access to the upper level(s). However, stairs have not been found in all Nuraghes, either because of their state of ruin or because it is effectively absent [5]. In more advanced Nuragic structures, a turreted walled enclosure (a sort of bulwark) was added and developed harmoniously around the main monument. Complex Nuraghes reached their greatest number and highest magnificence and elegance in the period between the Recent and Final Bronze Ages (about 1350 and 1100 BC); among them, about 40–50 are shown to be surrounded entirely by a wall with towers that are only one floor high. Therefore, the outer towers were under the control of the rampart and its main tower, which reached greater heights [6]. Nuraghes are found to be widespread throughout the Sardinian territory, from the coasts to the Gennargentu massif (central Sardinia), but they are more frequently found in the interior areas of the island. On the other hand, they tend to be less frequent in the areas in the middle of the Campidano plain (south-central Sardinia) and in the today's wooded reliefs [6]. In the present work, we studied a complex Nuraghe located in the territory of south Sardinia (Sa Domu e S'Orku in Domusnovas). This Nuraghe was already characterized in the second half of the 1800s [10,11] and described in more detail in subsequent work presented in the 1960s and 1980s [1]. This monument consists of a main tower with a bastion with polylobate development and a front courtyard. The main structure is included within a discontinuous wall with five perimeter towers (Figure 1). The masonry work is of the polyhedral type, with stone blocks arranged tendentially in oblique rows, forming a so-called "megalithic matrix" [1]. The masonry structures of the Nuragic complex consist of a double stone face of irregular (unworked or slightly hewed) blocks that show dimensional variability ranging from just under a meter to two meters and decreasing from the bottom to the top. The interior space of the masonry is occupied by rubble stone mixed with loose material of varying grain size. The internal tholos chambers of the towers have access corridors, niches, and, in the case of the main tower, a spiral staircase (currently not accessible because it is occluded by collapses) that led to the upper level [1].

Figure 1. The archaeological site of Nuraghe Sa Domu e S'Orku, Domusnovas. The images highlight the texture and complexity of the monument's forms and surfaces. (**a**,**b**) aerial views; (**c**) view from the south; (**d**) view from the north; (**e**) main tower entrance; (**f**) west court and northern tower.

In this work, we have implemented and tested 3D modelling integrated with the geolithological investigation of the area to improve the characterization of the Nuragic site. This study focused particularly on the geometric and dimensional characterization of the Nuragic structures still in place and on the typological definition of the materials used for their construction, within the geolithological context of the photogrammetric survey area and its surroundings. The effectiveness of this integrated approach has already been highlighted at the scale of the archaeological artifact. Some authors report schematic references of the architectural proportions of Nuraghes, reporting the main measured quantities and significant dimensional parameters used to describe and compare structures and to understand the variety of Nuragic architecture in relation to building types and techniques [10]. Dimensional analyses of these monuments are often presented in descriptive and qualitative terms, without a systematic study of the relationships and proportions between architectural elements [12]. In the present study, the geometric parameters of the Nuraghe were surveyed and measured in a systematic way, highlighting, in particular, the importance of carrying out an analysis of the unevenness of the terrain, which should also be reported below the architectural structure (obtained by correlating the RTK-GNSS topographic survey data with those obtained from the 3D photogrammetric models). This is in order to obtain the definition of the original elevation of the accesses to the towers and the walled enclosure by estimating the thicknesses of the collapses within the structures. These survey methods are also used to carry out the measurements and analysis of the proportions of the collapsed or partially buried parts as well.

3D modelling based on Structure-from-Motion Multi-View Stereo (SfM-MVS) techniques particularly in recent years has had a significant uptake in the field of digital documentation and the mapping of cultural heritage sites [13–16]. Numerous studies have shown that SfM-MVS photogrammetry, by optimizing the acquisition phase, can achieve levels of accuracy and completeness of 3D models comparable to those achievable through scanning (LiDAR) and laser scanning (TLS), presenting several advantages in terms of cost, the speed of data acquisition, processing, and the quality of both geometric and photorealistic 3D models [14,17]. Photogrammetry allows for operational flexibility with acquisition from multiple viewpoints by limiting the shadow areas and blind spots generated by occlusions, which are typical of laser scanner surveys. Topographic mapping at different scales is usually based on the use of a laser total station (TS), global navigation satellite system (GNSS), terrestrial laser scanning (TLS), and LiDAR scanning [18]. In areas with special morphological complexities, mapping is increasingly being performed using UAV systems equipped with full-frame DSLR and APS-C photographic sensors using SfM-MVS photogrammetric techniques. Using different approaches to SfM-MVS modeling, it is possible to reconstruct 3D models of an object based on the extraction of data obtained from sequences of 2D photographs taken with overlapping acquisitions, even using photos captured with uncalibrated cameras by varying the point of view angle [17–19]. In areas with particular morphological complexities and brightness problems, mapping is performed by nadiral and oblique image acquisition using UAV equipped with full-frame or APS-C cameras [17,18,20–22]. Large-scale systematic deformation can be reduced with the use of a network of ground control points (GCPs) [23]. Several authors have analyzed the influence of image overlap and the number and distribution of GCPs used for georeferencing 3D models as key aspects with respect to the accuracy of 3D reconstruction. In addressing the issues of the 3D modelling of cultural heritage sites, several authors have highlighted the advantages of SfM-MVS modeling also integrated with diagnostic material analysis techniques [15,18,23–25]. Photogrammetric surveys based on the use of UAV platforms are documented in the literature for the creation of digital models that, thanks to the use of various photogrammetric software available on the market, allow the generation of 3D models, DTM and DSM models, ortho-photos, and high-resolution topographic maps of artefacts and structures in archaeological and cultural heritage fields (but also for topographic mapping, monitoring, and the analysis of the degradation of monuments and structures for geomorphological analysis) [18,23,26–28]. At present, 3D recordings of prehistoric archaeological monuments in Sardinia (Italy) have only been carried out for a few cases. Among them, it is important to mention the Nuraghe Oes project [29], which provided a complete documentation of the state of conservation of a complex Nuraghe, adding to the archaeological knowledge about the nature of the monument, using several integrated geomatic techniques [29]. The precision of 3D SfM-MVS models processed from digital photogrammetric images depends on several factors such as the quality characteristics of the sensor, the lens focal length, the flight planning parameters of the UAV, the repeatability of the acquisition patterns, the orientation of the camera, the final quality of the images, the topographical and morphological characteristics of the area to be surveyed, the light conditions, the type and quality of the photogrammetric processing software used, as well as the distribution and quantity of ground control points and check points (GCPs and CPs) [27]. For applications requiring a greater precision of structures and architectural elements representation, the processing of the SfM-MVS 3D model can be implemented through an additional photogrammetric survey with a lower GSD (cm/pixel) integrated with LiDAR scanning and terrestrial TLS acquisitions to achieve greater levels of completeness and detail of 3D models especially in particularly complex structures [13,30]. The 3D models produced with these processing techniques enable the noninvasive metrological analysis of structures correlated with building materials' characterization, which can be applied in archaeology, conservation, and architecture fields. Such models can be used as basic tools for the planning of restoration work, maintenance scheduling, and documentation during archaeological excavations. They also allow for the creation of virtual reconstructions of partially collapsed or buried structures, providing useful insights for scholars about their original appearance. In addition, 3D reconstructions are effective for developing virtual and augmented reality (VR/AR) experiences, becoming additional tools for the enhancement and enjoyment of archaeological heritage even remotely through social media. Modelling techniques based on 3D SfM-MVS reconstruction can provide levels of accuracy comparable to those achievable with LiDAR scanning and laser scanner (TLS) acquisition, also providing an important contribution to digital documentation for the preservation of cultural heritage projects. These laser surveying techniques can be complementary to photogrammetric processing for the analysis and understanding of monuments, meeting versatility and accuracy requirements that allow for their integration into archaeological heritage enhancement studies [17,30].

2. Materials and Methods

2.1. Archaeological Site and Geographical Location

The archaeological site of the Nuraghe Sa Domu e S'Orku is located in the territory of the Municipality of Domusnovas (SU) and has a total extension of 26,000 m^2 (Figure 2). The photogrammetric survey area is geographically identified by the following vertices: 4,352,960.45 N, 1,468,931.67 E and 4,352,670.45 N, 1,469,121.67 E (Monte Mario/Italy Zone 1, EPSG:3003), corresponding to a total of about 0.0551 km^2 with a maximum elevation difference of about 17 m ($155 \leq Z \leq 138$ m). In this area, we can find the main monument that covers a total area of 1591 m^2, consisting of the Nuraghe with a bastion and central main tower (called the keep) surrounded by a wall equipped with five perimeter towers. Around the Nuraghe, there are several huts' remains, piles of stone blocks, and masonry remains, with a total area of about 6400 m^2. At the site, we distinguished a monument survey area with an extent of 80 × 80 m and a topographic mapping area extended to the entire archaeological site of 190 × 290 m. The dense vegetation cover, present at the time of the study, was removed as far as possible from the monument but remained scattered throughout the rest of the archaeological area. For the photogrammetric survey, the monument was divided into exterior and interior scenes. The size of the exterior scene was about 80 × 80 m and developed with a difference in elevation of about 15 m ($155 \leq Z \leq 140$ m). The interior scene includes accessible tholos chambers, their entrances, and corridors, where the maximum capture distance was less than 5 m.

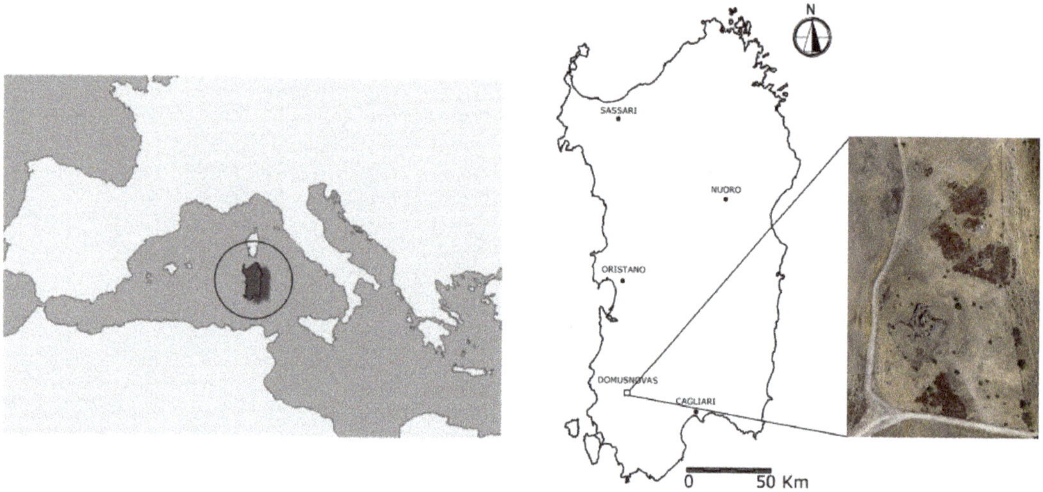

Figure 2. Geographical location of the archaeological site Sa Domu e S'Orku, Domusnovas.

2.2. Survey and Modelling Workflow

The adopted 3D surveying and modelling method was based on a workflow that involved defining the level of accuracy and visual quality required by the final 3D model. This depended on the minimum detail of the architectural elements of interest to be reproduced and the scale of the monument. The photogrammetric survey was performed with a UAV equipped with an APS-C sensor and developed following both nadiral parallel photogrammetric patterns and oblique converging shots. The UAV survey was integrated with ground details for partially covered and interior areas, following a scheme for parallel image sequences, which was acquired with normal and oblique axes to ensure the best base-to-height (B/H) ratio in the different capture situations [20,28]. The reconstruction of the 3D SfM-MVS model was carried out with reality capture software (RC) Version 1.3.1.117316 [31], subsequently simplified into sub-regions for visual and dimensional analysis at different scales. The quality validation of the 3D model is a key aspect. The estimation of errors in the georeferencing of shapes in terms of the 3D model's rigidity are verified by measurements on GCPs and CPs materialized using coded targets and measured using RTK-GNSS [17,30,32–34]. Surface accuracy can be estimated through comparison with benchmarks from the laser scanner survey (TLS) or the repeatability of the SfM-MVS process with the same APS-C sensor or full-frame sensor, along with the evaluation of completeness and the visual quality of textures [27,28,35].

The 3D SfM-MVS model obtained using the proposed workflow can achieve a level of accuracy consistent with the purposes of geometric and dimensional documentation and analysis at the scale of the archaeological monument. In addition to the 3D models, DTM and DSM, ortho-photos, and topographic maps, ortho-views are processed on characteristic sections and profiles of the monument so that measurements and comparisons can be made at the local scale. The methodology used for surveying and processing the 3D SfM-MVS model can be repeated systematically for other structures for typological and dimensional comparison.

2.3. UAV Survey and Photogrammetric Method

UAV flight planning for the aero-photogrammetric survey was carried out using Mikrokopter-Tool software, version 2.22 [36]. The survey was carried out according to flight plans with nadiral parallel shots and converging shots following circular orbits in accordance with the established optimal GSD, taking into account the variation in terrain and monument elevation. The value of the GSD for the oblique converging images was defined according to the average calculated over the overall values obtained, which considers the ground footprints of the camera image taken at the minimum and maximum distance from the object [20]. In-flight images were acquired with a Mikrokopter quadcopter [36], equipped with a Ricoh GR II camera (Table 1).

Table 1. UAV system and camera specifications.

UAV Parameters	Value	Camera Parameters	Value
Type	ARF-Mikrokopter	Sensor type	APS-C
Rotors	Quadro	Resolution (Mpix)	16.1
Size W × L × H (cm)	75 × 75 × 40	Focal length (mm)	18.3
Voltage (V)	14.8	Focal length equivalent (mm)	29.0
Battery (Mha)	LiPO 7200	Length sensor (mm)	15.6
MTOW (Kg)	2.8	Width sensor (mm)	23.7
Max flight time (min)	15	Pixel size (micron)	3.9
GPS navigation	U-BLOX MTK	Length frame (pix)	3264
Camera sensor	APS-C	Width frame (pix)	4928
Spectral range	RGB	Shot	Intervalometer

The nadiral flights over the archaeological area were oriented in the north–south direction and performed at altitudes of 30 m and 50 m, achieving coverage areas of 6400 m^2 and 55,100 m^2, respectively. The convergent shots were performed over the monument following circular flights orbits with the camera inclined at $15^\circ \leq \alpha \leq 60^\circ$ at different altitudes of $10 \leq Q \leq 45$ m. During flight in the real operational scenario, the camera altitude and pointing direction are always subject to some degree of variability from the theoretical planned flight lines (camera in nadiral or convergent orientation). This aspect generates a random noise component related to small variations in camera pointing direction (due to UAV roll, pitch, and yaw movements) or distance (altitude) variations related to terrain slope trends [23]. Converging image acquisition according to a circular orbit is shown in several studies to be particularly effective in reducing doming error [23]. Flight plans along parallel lines were programmed with an 85% overlap and 75% sidelap, set in accordance with the parameters of the camera used (sensor size, focal length, and flight altitude). At flight altitudes of 30 m and 50 m, B/H values were within the allowable range of <0.30. The same applied for the survey with circular orbits at altitudes in the range of $10 \leq Q \leq 35$ m with B/H values of <0.25. The camera was set in aperture priority mode, fixed to f/8 for nadir images and f/5.6 for oblique images and with a shutter speed of 1/1000 s. All images were captured in RAW format with ISO set to 250. To complement the UAV shots, images from the ground were also acquired with the same APS-C sensor. The use of parallel and oblique images is equally valid for ground-based image collection with consistent B/H ratio values and camera-to-object distance in the range of $1 \leq D \leq 5$ m [23,28]. When the survey targets indoor scenes and confined environments or transition zones between different scenes with illumination variability, image shooting becomes problematic in terms of acquisition mode and data quality. In such scenes, the field of view (FOV) of each photo turns out to be limited. This aspect affects the B/H ratios, whose correct setting is the basis for the photogrammetric accuracy expected with the overall specifications defined in the survey. The total number of images acquired in flight was 2128, while 1331 photos were taken from the ground, for a total of 3459. The average GSD in the survey area is 1.00 cm/pixel, and the average GSD in the monument area is 0.53 cm/pixel.

The total time to complete all acquisitions was 4 h, from UAV setup to the end of the survey. The RTK-GNSS surveys of GCPs and CPs control points, including ground points, were collected on geological outcrops. RAW images were post-processed to improve their quality by adjusting white balance, color correction, noise reduction, and exposure compensation parameters. Images acquired during the photogrammetric survey were processed using reality capture software [31] based on Structure-from-Motion Multi-View Stereo (SfM-MVS). 3D photogrammetric reconstruction was based on defining a pattern of a network of images with baselines, which must ensure adequate levels of overlap. This aspect is influenced by the resolution of the images, the angle of photographic capture, and the mutual distance between images, and it influences the quality of 3D reconstruction. Scanning based on SfM-MVS allows dense point clouds to be obtained with the use of uncalibrated perspective and spherical digital cameras with very good quality color data. In our case, we used a specific digital camera with an APS-C sensor that has a dynamic range value of 13.7 EV (>12EV high-quality limit) and excellent noise performance values at high and low sensitivity and color depth. The shape and size characteristics of the monument for both outdoor and indoor scenes required the combined use of different UAV and ground-based image acquisition techniques and materialized GCPs and CPs control points with encoded targets. The distribution of GCPs and CPs was based on the topography terrain characteristics of the area and those of the still outcropping structures of the monument. The identification of GCPs is related to image resolution and GSD. GCPs can be identified by encoded targets but also on natural objects that must be identifiable in images with resolution consistent with the GSD value used [26]. Several models of codified targets have been proposed in the literature, which must be clearly distinguished from each other and have a robust disambiguation with environmental characteristics. Ground-coded targets were accurately identified through automated procedures in the photogrammetric georeferencing process of the project coordinate system. In general, coded targets can be classified into three categories: 1D targets, planar targets, and 3D targets [33]. In our study, planar targets were chosen with patterns of concentric circles and black circular dots of the Schneider type and black and white patterns arranged within square frames, generated through QR codes of the AprilTags type [34]. Schneider targets were used for nadiral and oblique converging surveys performed with UAVs and AprilTags targets for the survey of entrances and internal parts of the monument. Schneider-type targets have a simple geometry that makes them more robust in low-light and low-resolution conditions than targets with more complex patterns. Moreover, the perspective deformation of the concentric shape means that the projection of the circles into ellipses with increasing distance does not compromise recognition in space [33,34]. AprilTags targets are designed to be easily identified and decoded by computer vision algorithms due to the strong contrast and well-defined edges that allow robust detection even in limited lighting conditions. To correctly reconstruct the relationship between the 2D coordinates of each image and the coordinates in 3D space, the automatic identification of the measurement points, on the basis of the points of the image, is essential [32,33]. The targets used in the present study have dimensions of 29 $\times$ 29 cm and 15 $\times$ 15 cm and were printed on rigid supports carefully positioned on the points on the ground. With the use of these targets, the process was accelerated, improving the accuracy and reliability of the measurements on the 3D model. GCPs and CPs have been materialized both as reference vertices, through coded artificial targets, and as natural feature points on the ground or structure in question (Figure 3). The network of GCPs and CPs points was measured with GNSS (global navigation satellite system) real-time kinematic (RTK) techniques using a Topcon Corporation (Tokyo, Japan) Legacy-Hyper V5 receiver in RTK mode. GCPs and CPs were measured using a rover and GPS base station. The rover antenna was set to PDOP = 4.0, and horizontal and vertical precision tolerances were set to 0.010 m + 1 ppm and 0.015 m + 1 ppm, respectively, while the base was set to have a 15° elevation mask. The base station used for RTK-GNSS surveying recorded for a time of 1 h on a point of known coordinates. The accuracy level of the base station was 0.01 m in the vertical and horizontal components. With the rover antenna, data were recorded for each GCP and CP for 5 min. The accuracy of the kinematic method allowed for centimeter-level accuracy.

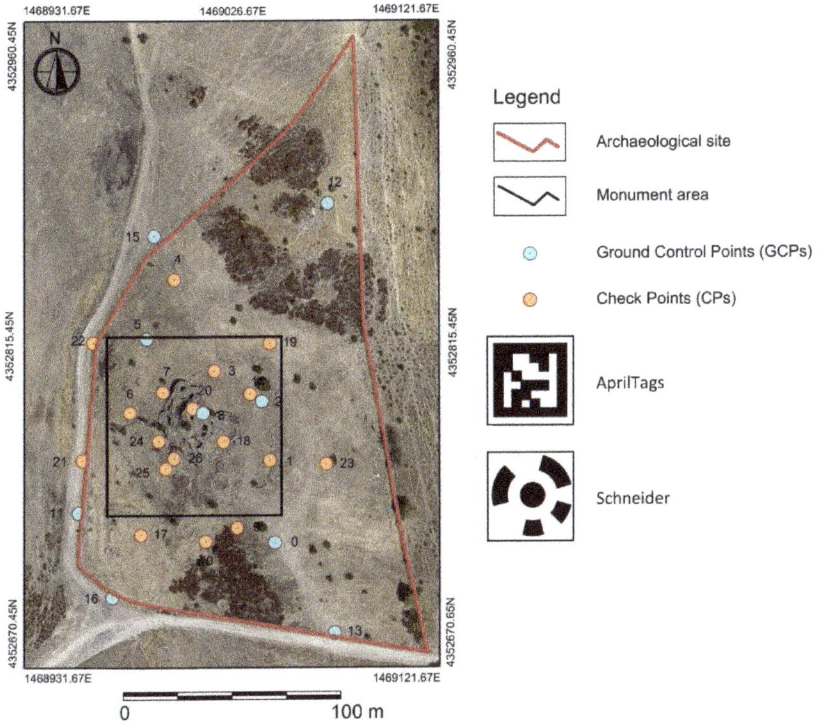

Figure 3. Spatial distribution of ground control points (GCPs) and check points (CPs) and target types used in photogrammetric survey.

Using a series of oblique convergent images, it is possible to reduce systematic error in stereo pairs acquired with linear UAV flight patterns. Therefore, by integrating nadiral acquisition with convergent acquisition, as demonstrated by several studies, systematic deformation can be effectively reduced [23]. The parameter that most influences the accuracy of a photogrammetric model is the base-to-height ratio (B/H) [28]. For nadiral photogrammetry, the optimal B/H ratio can vary according to the geometry of the object or scene and is in the range of $0.2 \leq B/H \leq 0.4$; for convergent photogrammetry, the optimal B/H ratio, which can also vary according to the geometry at the shooting angle, is in the range of $0.1 \leq B/H \leq 0.5$ [24,25]. Several studies have shown that for vegetation-free soil types and grass-covered areas, measurements from multiple stereo pairs are quite consistent with an advantage to DSM accuracy. In the case of areas covered by high vegetation, the dispersion of the elevation data from stereo pairs is significantly greater. The quality of the elevation measurement is highly dependent on the B/H ratio and the type of ground cover. On flat areas of vegetation-free terrain, elevation assumptions from high B/H ratio stereo pairs are more consistent. On the other hand, in areas with vegetation, most assumptions of elevation from high B/H ratio stereo pairs are anomalous [37]. Elevation data from stereo pairs with lower B/H ratios generally better represent the ground surface and, importantly, contain significantly fewer outliers. This is because nearby images are more similar to each other and, therefore, searching for matches is more efficient. Elevation data from stereo pairs with the lowest B/H ratios allow for a first approximation of the reconstructed surface [37].

2.4. 3D Modelling

The reality capture software (RC) [31] uses an evolution of sophisticated algorithms for Multi-View Stereo Reconstruction (MVS) and allows for the management of large amounts of data to obtain high processing speeds, to guarantee reliability, and to extract highly structured surfaces, as demonstrated by several authors [17,24,25,38,39]. It allows for automatic classification through artificial intelligence (AI classification on DSM), editing interventions with semi-automatic procedures on 3D models and textures. AI classification can allow for the creation of ortho-views, cross-sections, and measurements extrapolated directly from 3D models. It enables the definition of standardized and repeatable workflows. The workstation used for processing was a QuadCore Intel Core i7-4820K (8 Mb Cache) 3700 MHz CPU with 64 GB of DDR3-1333 MHz RAM and an NVIDIA GTX 1080 Ti GPU with 11 GB GDDR5X. In RC, multiple RRs have been defined for each survey area to limit the reconstruction of 3D space. We have defined RR_1 for the survey area for topographic mapping with an extension of 190 × 290 × 32 m and RR_2 for the monument area with an extension of 80 × 80 × 18 m. The relief area of the RR_2 monument area has been divided, as described above, into external and internal scenes. A further RR_3 region with an extension of 48 × 45 × 18 m has been defined to elaborate the ortho-views and characteristic sections of the Nuraghe. The use of reconstruction regions makes it possible to optimize the photogrammetric calculation process and to focus dimensional and accuracy analyses at the local scale on parts of the monument of particular interest. Multiple RR reconstruction regions allow for the optimization of the reconstruction of specific parts of the scene, speeding up the photogrammetric process and improving the quality of processing. Each region can be moved, rotated, and resized via the RC widget. The reconstruction regions defined in the project can be exported and imported to other platforms (CAD, GIS) to generate 3D sub-models, ortho-views, and ortho-sections on specific areas of interest of the 3D model.

2.5. Accuracy Assessment

The level of accuracy and visual quality required in the final 3D model and the quality constraint required to validate the geometry of the SfM-MVS 3D model were established with a maximum allowable error of <=2.5 cm. For this purpose, the georeferencing accuracy was evaluated on the basis of the estimation of the X,Y,Z residuals of the GCPs calculated from the RTK-GNSS measurements. The rigidity of the 3D model influencing the propagation of errors in space was evaluated through the estimation of the errors of the Euclidean reciprocal distances of the GCPs and CPs with a comparison between the calculated and measured data. The error of coherence and geometric completeness of the surface reconstruction was estimated with a cloud-to-cloud distance (C2C) using Cloud Compare software (CC) Version 2.12.4 [40] through a comparison between the sample surfaces processed from images acquired with the same APS-C sensor as that used on the typical surfaces of the SfM-MVS 3D model. Finally, an analysis of the completeness and visual quality of the texture of the reconstructed surfaces was carried out. The positioning accuracy of a 3D model was evaluated by calculating the mean (MEAN), standard deviation (STD), mean absolute error (MAE), and root mean squared error (RMSE) of the residuals on the GCPs (9) and CPs (18) using the coordinates measured on the ground with the global positioning system (RTK-GNSS) and those calculated using the photogrammetric model [28,32]. To evaluate the global geometric rigidity of the SfM-MVS 3D, the Euclidean distances of a 3D Delaunay network defined using GCPs and CPs were compared with those measured using RTK-GNSS at the site for the estimation of the RMSE value [17]. The evaluation of the quality and completeness of the SFM-MVS 3D model at the local level was based on the assessment of the error of the mesh vertices with a direct comparison between a generated point cloud and a ground truth reference point cloud. TLS scanning is widely used in the documentation of architectural heritage and allows data to be acquired with millimeter-level precision for 3D modelling. Several studies have verified the performance of SfM-MVS models obtained using full-frame DSLR sensors compared to a reference "ground truth" benchmark acquired using a terrestrial laser scanner (TLS). For full-frame DSLR sensors, C2C errors < 0.5 cm with 95% confidence and RMSE on Euclidean distances of <0.8 cm compared to the laser scanner [17] were obtained. For the APS-C sensor, we repeated the tests performed on surface types with "ground truth" benchmark comparison with the Faro Scene 350 at distances of 5 and 10 m. This laser scanner (Class 1, λ = 1550 nm) has a measurement error of $\pm$1 mm and an accuracy in detecting 3D positions of 2 mm at $\pm$ 10 m and 3.5 mm at 25 m; for longer distances, the accuracy decreases in increments of 0.1 mm per meter. C2C errors of <0.7 cm with 95% confidence and RMSEs on Euclidean distances of <0.9 cm were obtained. The C2C error is about 1/1000 of the maximum measured width, and it shows good overall accuracy of the photogrammetry model compared to the reference laser scanner. In our study, the quality of the reconstruction in terms of surface deviation was evaluated on typical surfaces of the Nuraghe, using our APS-C sensor mounted on tripods and telescopic masts, to obtain a comparison based on a benchmark "ground truth" positioned at 5 m from walls. The aim was to verify whether the quality of the reconstruction from the APS-C camera could be consistent in terms of geometry and texture quality with the requirements of the proposed photogrammetric workflow. The use of the same sensor has the advantage of having the same internal acquisition geometry, the absence of systematic distortions due to the use of different sensors, and a direct comparison between the "ground truth" and the SfM-MVS 3D model. Each benchmark was defined by ratio of 1:10 between the GSD of the photogrammetric survey and that of the "ground truth", which is largely sufficient to perform a reliable analysis of geometric deviations in surfaces. In case the GSD survey for some walls is lower than or equal to the GSD benchmark, it is possible to exclude these areas from the C2C validation, as the resolution of the photogrammetric model is already higher than or equivalent to the "ground truth". The surfaces selected as typical have covered a variety of shapes of walls, materials, and textures present, so that a meaningful assessment of model completeness could be obtained. The datasets acquired for each typical surface were aligned in the SFM phase with the survey data to then calculate the individual "ground truth" MVS models for the C2C validation. The following is given as an example of a photogrammetric benchmark performed on a typical surface represented by the external face of the M tower, which develops on the XY and XZ planes, delimited by the RR_4 reconstruction region with a size of 12 $\times$ 4 $\times$ 5 m. This process employed the use of a tripod/mast, remote control shooting, constant light conditions, and specific base/height (B/H) ratios, ensuring optimal baseline configuration. The capture GSD was 0.13 cm/pixel (camera–object distance = 5 m). The M1 "ground truth" model was calculated in RC with 1 cm spacing. In the same way, the M2 "flight and ground survey" model was calculated in RC, delimited by the RR4 with a spacing of 1 cm with an average GSD from the overall survey equal to 0.54 cm/pixel (camera–object distance settings varying in the ranges of $5 \leq D \leq 50$ m for the external scene and $1 \leq D \leq 5$ m for the internal and access scenes). The validation of the M2 model was performed using the cloud-to-cloud distance C2C (M1–M2) comparison based on the Hausdorff distance between point clouds with 99% outlier elimination. This validation also included an analysis of the completeness and visual quality of the texture of the reconstructed surfaces.

2.6. Characterization of Building Materials and Geoarchaeological Investigation

A typological evaluation of the nature of building materials used for the construction of the monument was based on a sampling of the stone ashlars and the filling material of the masonry structures, carried out in compliance with the visual integrity of the Nuraghe. The sampling was also extended to the geological outcrops present in the archaeological area. On these outcrops, a survey of the layering parameters was also carried out to correctly report the trend of the geological substrate in the elaboration of the geological map of the archaeological site. For the sampling, to define the compositional characteristics of the investigated materials, XRD mineralogical analysis was performed using a Rigaku Corporation (Tokyo, Japan) Geigerflex D-Max/B X-ray diffractometer with a Cu anode, operating at

30 kV and 30 mA. On the basis of the results obtained in this phase, a geoarchaeological survey was planned beyond the limits of the archaeological site in order to verify the presence of geomaterials compatible with those used for the construction of the Nuraghe. During this survey, several stone blocks present in the slope deposits around Mt. Acqua and from alluvial deposits in the Rio San Giovanni valley were sampled and analyzed. Finally, the mineralogical characteristics of the monument sampling were compared with those of the geomaterials present in the investigated area to verify compatibility and identify possible places that supplied the stone used in the Nuraghe.

3. Results

3.1. Geological Investigation on Building Materials

The archaeological site is an area that is mainly characterized by geological outcrops attributed to the Paleozoic basement of Sardinia (Figures 4 and 5). In particular, the monument rests on metamorphic outcrops consisting of metasiltstones, metapelites, and subordinated quartzites, referring to the Medau Murtas Member of the Medau Argentu Formation (middle-upper Ordovician ?; [41]).

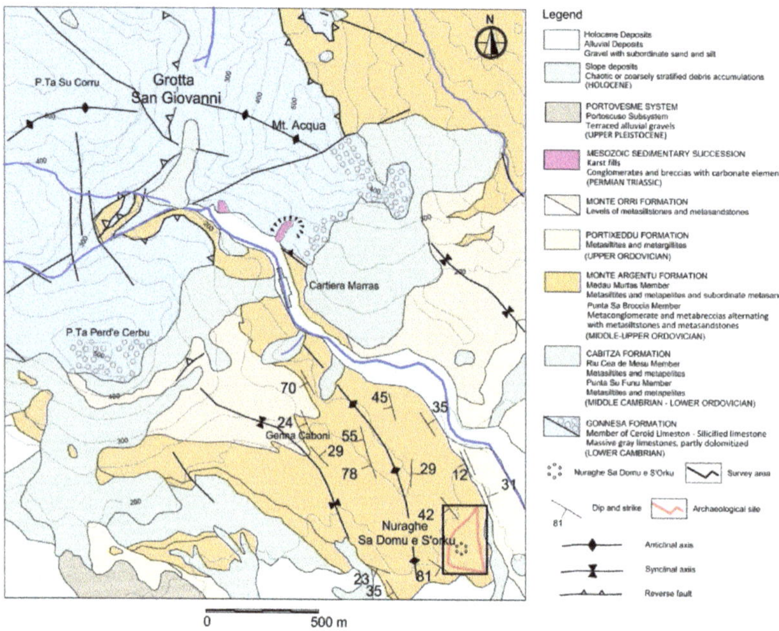

Figure 4. Geolithological sketch map of the studied area.

Outside the limits of the actual archaeological site, in the northern direction, the territory is dominated by limestone outcrops attributed to the Calcare Ceroide Member of the Gonnesa Formation, which constitutes the reliefs of Mount Acqua, Punta Su Corru, and Punta Perd'e Cerbu (Lower Cambrian; [41]). A substantial portion of these outcrops has undergone a profound phenomenon of silicification that has led to the complete transformation of the original limestones. This silicification has particularly affected the southeastern side of Mt. Acqua and the southern part of Punta Perd'e Cerbu. Substantial deposits of these silicified materials were found on the slopes of Mount Acqua and in the riverbed of the Rio San Giovanni in the form of flat or cuneiform parallelepiped blocks, variously elaborated, ranging in size from 0.5 m to 2 m (Figure 6).

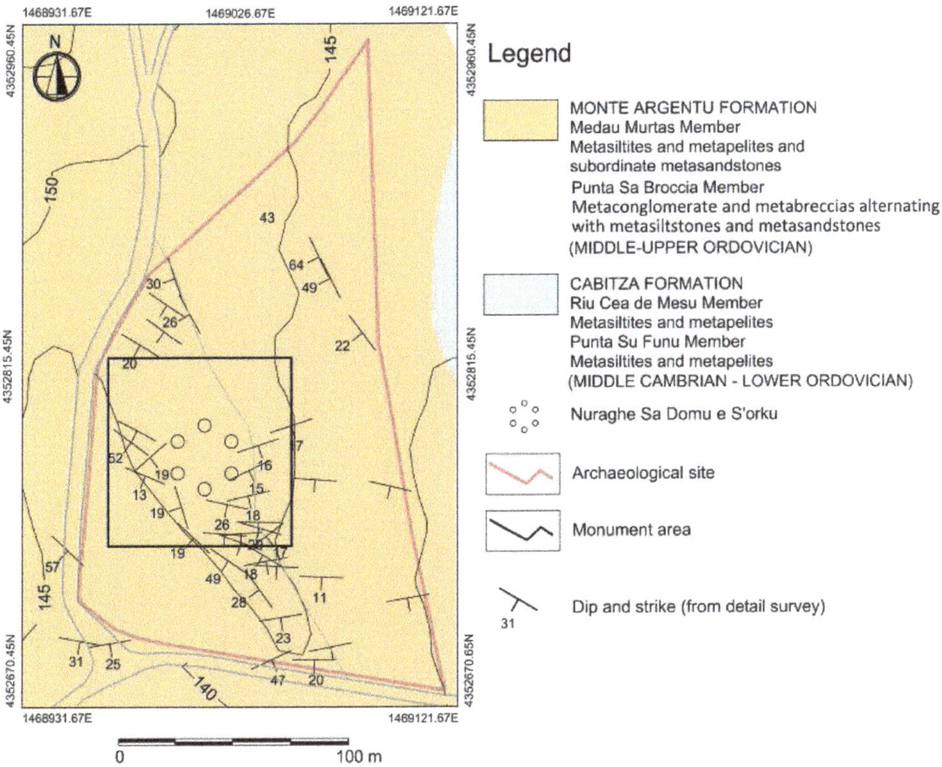

Figure 5. Geological sketch map of archaeological site. Ground points measured by RTK-GNSS on geological outcrops to estimate the elevation at the base of the Nuraghe.

Figure 6. (**a**) Blocks deposited along the valley of the Rio San Giovanni containing breccias; (**b**) partially elaborated silicified blocks coming from the slopes of Mt. Acqua.

As described in the previous section, the characteristics of building materials used in the structures of the Nuragic site were defined starting from their typological characterization based on macroscopic observations and then more deeply through XRD analysis (Figures 7 and 8). This highlighted the prevalent use of fully silicified lithologies. Given their composition, stone ashlars are extremely tough and appear unshaped or slightly hewn and laid dry (without mortar) according to a complex polygonal pattern of joints. Alternatively, there are small- and medium-sized blocks of metapelite and metasiltstone, used mainly to fill the voids and hold the larger elements of the masonry in place. In the partially collapsed structures, it is possible to observe the masonry construction technique, characterized by two faces, consisting of blocks inclined toward the inside and opposed to each other, whose major axis is at least 2/3 inside the masonry itself. The dry masonry always contains a variable amount of compacted filling material between the two facings, consisting of stone fragments with an average size in the decimetric range, immersed in a matrix with gravelly, sandy grain size. The results of the analyses on the samples taken during the geoarchaeological investigation showed a marked similarity in compositional and morphological characteristics between the stone blocks in place in the wall structures of the Nuraghe and the coarser elements of the deposits present in large quantities in the valley of the Rio San Giovanni and on the slopes of Mt. Acqua.

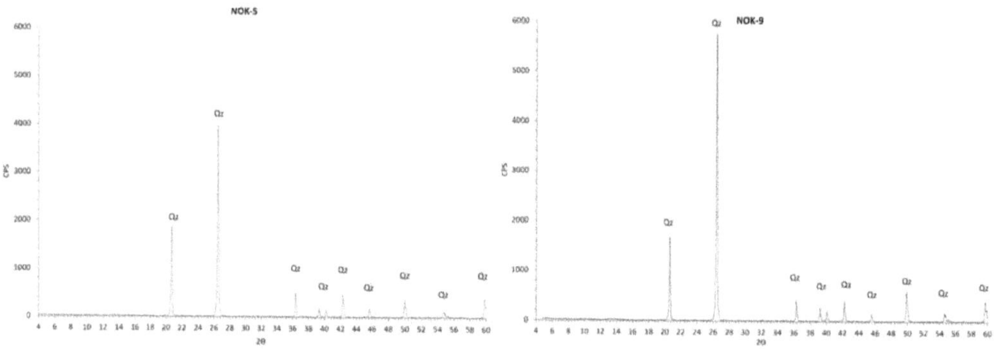

Figure 7. XRD analysis of stone blocks from the Nuraghe and comparable geomaterials of the investigated area. NOK-5 sample from main tower wall; NOK-9 sample of silicified stone block from stratified slope deposits of Mount Acqua. Qz = quartz.

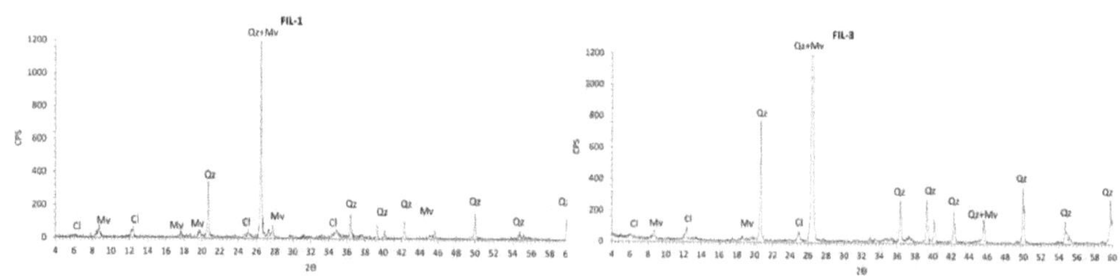

Figure 8. XRD analysis of filling material from the inside part of the masonry and compatible material from the riverbed of the Rio San Giovanni. FIL-1 sample from the inside of the rampart; FIL-3 sample from riverbed deposit. Qz = quartz; Mu = muscovite; Cl = chlorite.

The filling material inside the walls is also compatible with local colluvial deposits and sandy and gravelly deposits along the valley and in the bed of the Rio San Giovanni, derived from the erosion and transport of metamorphic basement rocks. Among the elements of the filling material in the walls of the Nuraghe, we found large amounts of refined metasiltstone and metapelite pebbles immersed in a sandy–silty fraction that show the same mineralogical composition as these metamorphic lithologies. The presence of refined pebbles leads to the reasonable hypothesis that these sandy–gravelly fractions came from the river deposits of the Rio San Giovanni.

The definition of the typological characteristics of the geological materials used for the construction of the Nuraghe and the attribution of the areas of origin are decisive in enriching the knowledge of the technological level reached by the Nuragic civilization.

The contribution of the laboratory analyses and the geological survey of the area confirm these data objectively, allowing for the verification of the preliminary classification of building materials performed on an autoptic basis. In addition, it has been pointed out that the mineral–petrographic characteristics of the stone material denote its particular hardness and the objective difficulty to model the blocks into regular shapes with the tools available at the time. However, this apparent limitation did not prevent the Nuragic builders from creating a complex monument, characterized by the careful selection of the shapes and sizes of blocks found in natural deposits. The decision to use predominantly silicified materials (consisting, as demonstrated, of pure quartz) for the Nuraghe blocks, rather than the limestones widely available in the region, is likely related both to their great availability in the proximities of protohistoric settlement and to their greater durability and resistance over time.

3.2. 3D Modelling of the Nuraghe

The 3D reconstruction workflow used in the reality capture software (RC) included the image alignment based on GCPs and CPs, SFM sparse point cloud reconstruction, and dense MVS (Table 2).

Table 2. Advanced alignment parameters in reality capture.

Alignment Parameters	Value
Alignment mode	High
Max features per Mpix	10,000
Max features per images	40,000
Detector sensitivity	High
Preselector feature	20,000
Image download factor for depth maps	1
Max feature reprojection error (pix)	1
Lens distortion model	K + Brown 3 with tangential 2
Alignment time	02 h:19 m:25 s
Feature detection time	00 h:26 m:41 s
Registration time	01 h:52 m:43 s
Maximal error (pix)	0.999999
Median reprojection error (pix)	0.336809
Mean reprojection error (pix)	0.381691
Camera count	3459
Point count	7,845,986
Control point used	9(GCPs) + 18(CPs)
Background thread priority	Normal

Next, the cleaning and topology of the 3D model, mesh texturing and simplification, 3D tiling, and the creation of sub-models for each defined reconstruction region were performed. Finally, the accuracy of the 3D model was evaluated and exported for further processing. In RC, the design coordinate reference system was EPSG:3003. For the SfM alignment process, the distortion model "K + Brown 3 with tangential 2" was used, and then the positions of the cameras were obtained and the sparse point cloud computed. In the subsequent phase, MVS computed the dense point cloud that was restricted to the reconstruction region RR_1 and RR_2, with a minimum spacing of 1 cm. RC reconstruction was performed in "High" mode, with an image resize factor for depth maps set to 1. The mesh model was calculated from the dense point cloud (Table 3). The model was cropped, simplified, re-textured, and exported as a point cloud in the .xyz and mesh format .obj to complete the dimensional survey and 3D visualization. The next phase of model simplification focused on consistently maintaining the original geometry in shape and size for each region of reconstruction RR model. This was performed in relation to the best ratio between the reduction in triangle area and the conservation of the quality of geometric representation. In RC, by adjusting the simplification parameters, it was possible to achieve a good compromise between the simplification and preservation of the original geometry, in order to preserve the mesh connectivity, the original topology, the edges, and the salient characteristics of the key vertex, avoiding even a partial collapse of regions. For validation, 3D metrics were used between key points, and cloud-to-cloud Hausdorff distance C2C comparison was used between original and simplified mesh. This phase resulted in simplified 3D models for each reconstruction region with a minimum edge length of >1 cm, consistent with the maximum constraint error. The reprocessing of the textures from the original models to the simplified ones has preserved the overall visual quality of the 3D reconstruction.

Table 3. Advanced reconstruction parameters in reality capture.

Reconstruction Parameters	Value		Value
Reconstruction region	RR_1		RR_2
Dimensions L × l (m)	290 × 190		80 × 80
Quality level		High	
Minimal distance between two vertices (m)		0.01	
Triangle count	372.8 M		109.6 M
Vertex count	187.0 M		55.0 M
Depth-map computation time	05 h:35 m:10 s		04 h:39 m:57 s
Meshing time	15 h:37 m:40 s		09 h:52 m:50 s
Overall processing time	21 h:12 m:50 s		14 h:32 m:47 s

The SfM-MVS 3D model of the monument allowed for the evaluation of the above-ground volumes to define the proportions of the collapsed and covered parts and to formulate hypotheses about the original dimensions of the Nuraghe. The 3D model reproduces a "Digital Twin" of a "complex Nuraghe", with a floor plan illustrating a main tower, a rampart with a polylobate structure surrounded entirely by walls and five perimeter towers. The representation scale of the 2D and 3D drawings was 1:100 for ortho-views and 1:500 for ortho-photos and DSM, with the maximum allowable error congruent with the desired level of detail <= 2.5 cm (Figures 9 and 10). This value was the quality constraint to validate the geometry of the SfM-MVS 3D model.

Figure 9. (**a**) Aerial photo of the Nuraghe (SW–NE); (**b**) view of the optimized 3D model of the Nuraghe in RC according to the reconstruction region defined for the monument area under survey (SW–NE); (**c**,**d**) views of the textured mesh of internal chamber and external walls; (**e**,**f**) example photos of internal chamber and external walls.

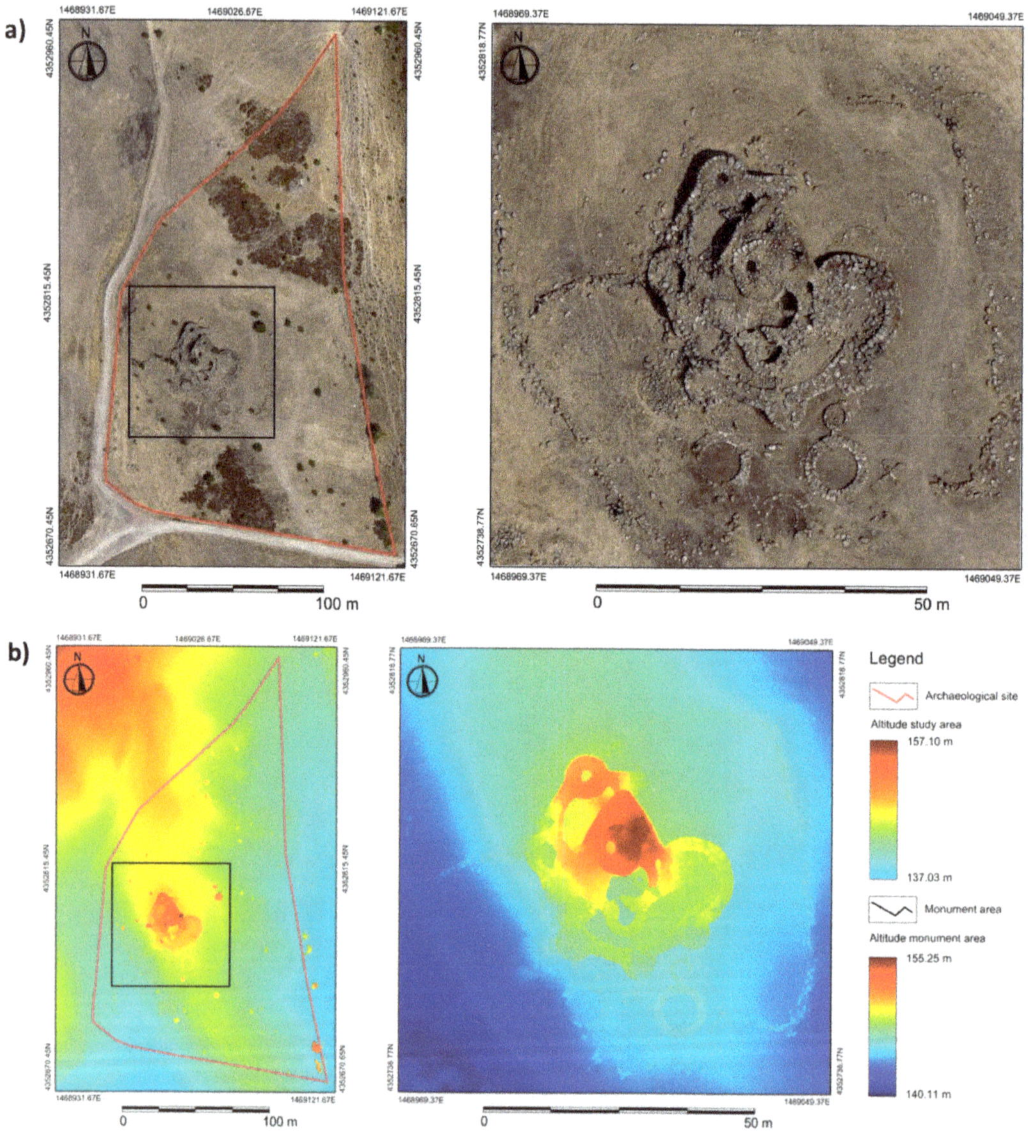

Figure 10. (**a**) Ortho-photo of photogrammetric survey area and monument area (left); (**b**) DSMs of archaeological site.

The hypothesis of the original proportions of the monument, based on previous observations of this Nuraghe and similar structures [1,3,8–11], was tested also through a purely geometric approach, based on the relationships between dimensions of diameters, angles, and profiles of the above-ground structures from the sections calculated from the SfM-MVS 3D model. The dimensional ratios and shapes typical of Nuragic structures were estimated, such as the height/diameter (H/D) ratio of the towers and their tholos chambers. Hypotheses were made to complete the profiles of the best fit of these chambers starting from characteristic points identified in the type sections. Proportionality and the integration of the missing parts were applied based on existing structures. The consistency of the estimated proportions with the geometrical constraints compatible with the still-accessible structures was verified. Particular attention was paid to the height of the access openings to the Nuragic towers and to the perimetral walls to define the lower levels of the entrances to complete the TIN of the rocky outcrops at the foundation of the Nuraghe. By processing in a CAD environment with parametric algorithms, it was possible to define a 3D Delaunay triangulation of the ground points detected on geological outcrops to define the share of the foundation base of the Nuraghe and to identify some 2D geometric descriptors of the structure from the characteristic sections. This allows for the reconstruction of 3D geometries for extrusion and the representation of the original dimensions and proportions of the monument based on the currently accessible measures of heights and wall thicknesses. Parametric processing with CAD on the 3D SfM-MVS model of the structures allowed for the definition of the original elementary volumes using interpolated lines and curves and the TIN model of the ground on which the monument rests as a basis. Using CAD tools for parametric 3D modelling, from basic 2D geometries, it was possible to obtain 3D models reconstructed according to the original proportions data. The maximum height for the towers was estimated starting from the constraints defined by the external basic dimensions and tholos chambers, the inclination of the external wall profiles, and the estimated shape and height of the internal elliptical paraboloid profile of the chamber. The strategy in use is based on the progressive calculation of the reduction upwards of the tower section and the wall thickness as a function of the inclinations of the two sides. In this way, it was possible to estimate the maximum height beyond which it is no longer possible to continue due to geometric limitations related to the wall thickness, the variability in the shape and dimensions of the blocks, and the maximum height of the inside profile. The maximum height of the tower was hypothesized based on the height constraint established according to the profile of the tholos chamber and the assumed thickness of the roofing material placed on its top. The construction solution adopted in the Nuraghe investigated involved the use of stone blocks that were irregular and polyhedral wedge-shaped, arranged to obtain walls that were perfectly aligned according to a precise inclination. The choice of wedge and flat shape is effective both for the towers of circular section, where the internal and external radius of curvature are concentric, and for open and closed polylines (as for the bastion and the perimetral walls), where it offers good adaptability in areas with a strong convergence of structures as a function of variability in wall thickness and inclination. Through the analysis of the 3D model SfM-MVS, all the information was obtained to formulate a geometric hypothesis that allows for the reconstruction of the missing parts, according to the construction rules typical of Nuragic structures. This approach allowed for the proposal of the maximum height of the towers, of the bastion, and of the perimetral walls and the evaluation of their reciprocal proportions. Area, volume, and other measurements on the 3D model were made directly in reality capture (RC) and in a CAD environment.

Studies in this field often report indications on the architectural proportions of Nuraghes, indicating the main sizes and significant dimensional parameters used to describe and compare structures and understand the variety of Nuragic architecture according to types and construction techniques. However, dimensional parameters are often presented in descriptive and qualitative terms, without a systematic analysis of ratios and proportions between architectural elements [12]. In this study, geometric parameters were systematically detected and measured. According to Vanzetti et al. [12], the main measured quantities and significant dimensional parameters to describe Nuragic structures are as follows:

- Height of the tower (H): a fundamental measure to quantify the overall dimensions of the Nuraghe. Variability from a few meters up to about 20 m.
- Diameter of the tower (D): indicates the width of the building base. Usually in the range of $10 \leq D \leq 20$ m in larger Nuraghes.
- Number of towers: indicates the degree of articulation of the structure in complex Nuraghes.
- Presence of walls: particularly in complex Nuraghes.

- Wall thickness: construction scheme with double side, with intramural space that can reach several meters in the most imposing Nuraghes, giving stability to the structure.
- Dimensions of the internal chambers: diameter (d) from 3 m up to about 6 m and height (h) from 3 m to about 8 m are key parameters to evaluate the organization of the interior spaces.
- Dimensions of openings and corridors: indicate the width of the passage points in the structure.
- Materials, dimensions, and shape of boulders: length, width, and thickness recurring in erratic boulders as well as their shape and physical characteristics in terms of workability are an index of construction capabilities.

Below is a report of the main dimensions measured in the Nuraghe studied. The towers indicated in Figures 11–13 with A, M, G, H, I, and L have a truncated cone shape with heights above ground between 2.10 and 7.00 m and total heights estimated between 5.50 and 14.30 m.

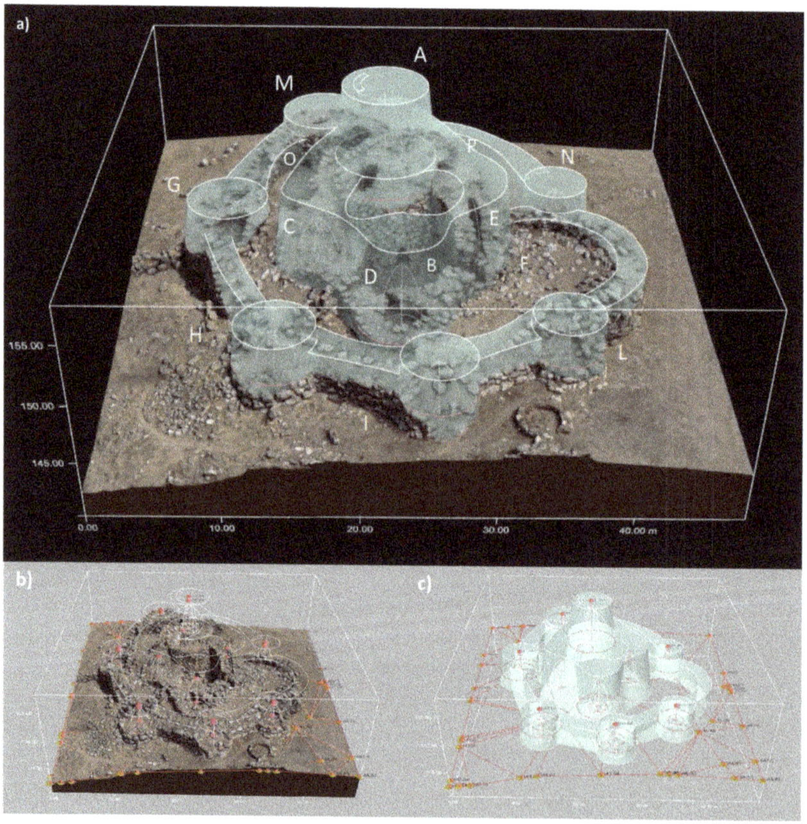

Figure 11. (**a–c**) 3D reconstruction of the Nuraghe with visualization of the missing parts (RR_3) obtained on the TIN foundation base of the Nuraghe. The different parts of the monument (from A to P) are described in the paragraph below. The approach used allowed for the estimation of the maximum height and the definition of the proportions between towers, ramparts, and perimeter walls.

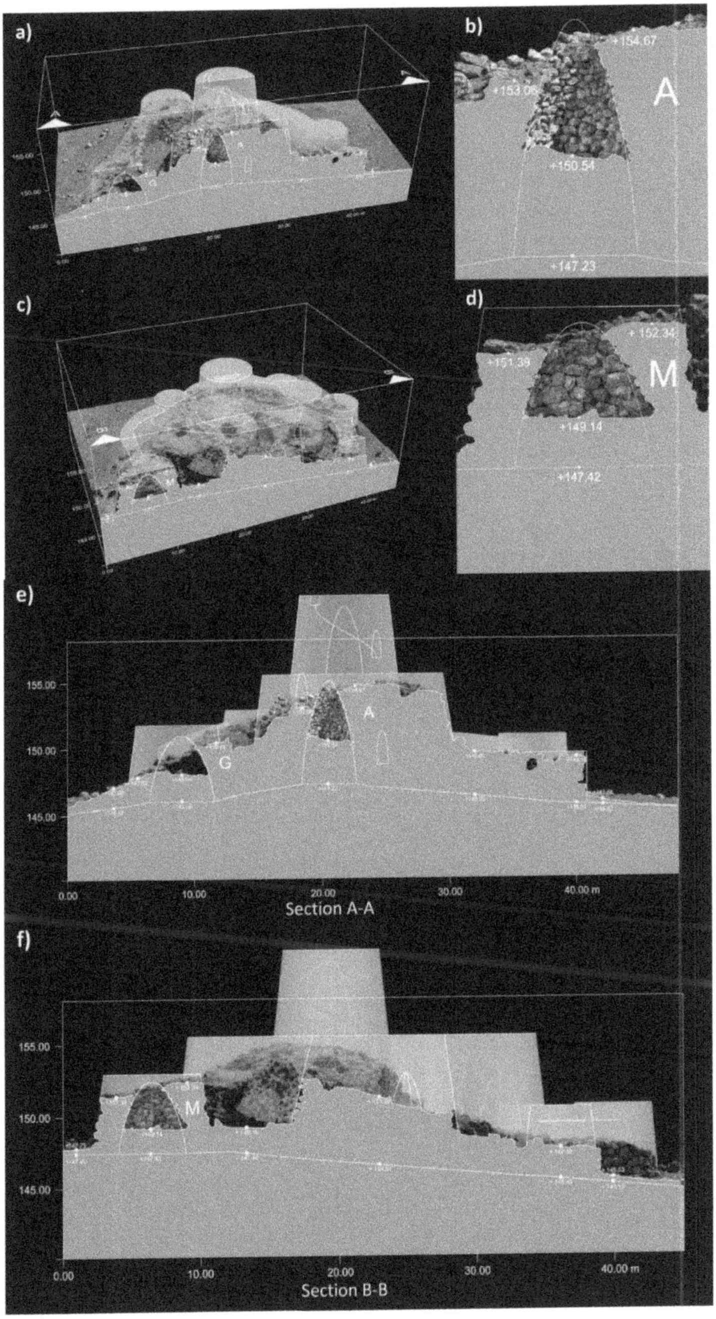

Figure 12. 3D model of the Nuraghe and hypothesis of reconstruction of the original dimensions of the monument. (**a**,**c**,**e**,**f**) Perspective and orthostatic view of sections A-A and B-B; (**b**,**d**) simplified 3D model: detail of the tholos of towers A and M.

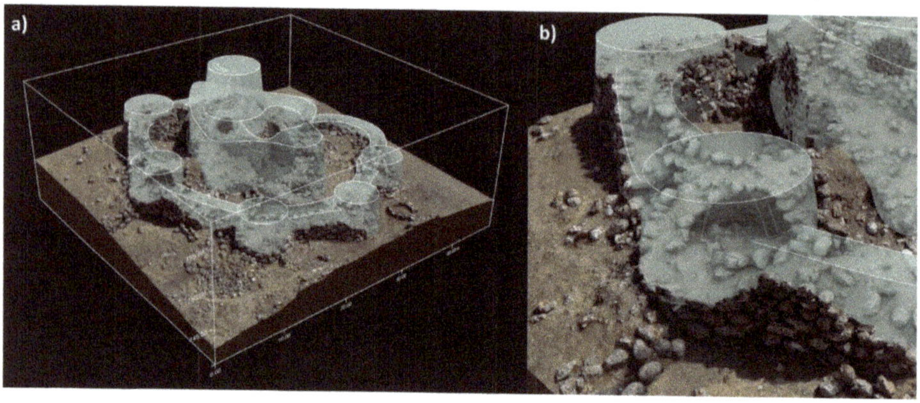

Figure 13. 3D model of the Nuraghe and reconstruction of the original proportions of the monument. (**a**) Perspective view; (**b**) detailed view.

The estimated base diameters vary from 6.10 to 10.20 m, and the top diameters vary from 5.40 to 7.00 m. The average slopes are between 4.0° and 8.65°. H/D ratios are between 0.63 and 1.40. The internal tholos chambers have circular or elliptical base shapes, with estimated heights of 4.84 to 7.80 m and average diameters of approximately 3.10 to 5.30 m. h/d ratios are in the range of 1–1.47. The estimated surface walls of the tholos chambers are elliptical paraboloid. The main tholos A has an elliptical base with axes of 5.30 m and 4.30 m and an estimated height of 7.80 m. At the upper level, the second tholos is presumed to have a circular base of 2.87 m in diameter and 5.25 m in height. The minimum thickness above the tholos chambers is established as >0.50m. The wall thickness, off-center location, and size of the tholos on both levels of Tower A are geometrically consistent with the hypothesis of the existence of a connecting staircase with helical development. This, as evidenced by previous studies [1], starts from the right side of the access corridor and likely reaches the top of the tower [10,11]. The ratios between the diameter of the tower and the diameter of the tholos chamber (D/d) are between 1.55 and 2.10. The axis of the access corridor to the main tower is oriented N154° E. Access exposure is in the SE quadrant. The maximum dimensions of the blocks of the load-bearing structures are in the order of 1.40–1.80 m in width, 0.70–1.00 m in height, and 0.90–1.20 m in thickness. The minimum dimensions are 0.40–0.60 m in width, 0.35–0.50 m in height, and 0.60–0.80 m in thickness. The tholos blocks have minimum dimensions of 0.4–0.5 m in width, 0.30–0.35 m in height, and 0.50–0.70 m in thickness. Tower N and part of the east perimeter wall were assumed to exist based on earlier indications found in other studies [1,10,11]. The rampart has a polylobate structure and a current height above the ground of 7.40 m, while the original height is estimated to be 8.40 m. The outer wall side has an average slope of 6.95°. Although not perfectly visible and measurable, two other possible chambers D and E are identified in the Nuraghe. A tholos with a circular base of 3.20 m diameter and a height of 5.65 m, defined according to an h/d ratio of 1.76, is assumed for chamber D. The tholos of chamber E has the same height as the previous one and an elliptical base with axes of 4.30 m and 3.10 m, from which an average h/d ratio of 1.31 is derived. The estimated surface wall of the tholos chambers is an elliptical paraboloid. The area of the internal courtyard B is approximately 30 m^2, while the footprint of the keep is 315 m^2. The lateral tholos chamber C to the west was hypothesized on the basis of estimates of the proportions of the chambers and previous indications made by other studies [1,10,11]. The maximum dimensions of the blocks at the base are 0.90 m width, 0.70 m height, and 0.95 m thickness, and the minimum dimensions are 0.60 m width, 0.40 m height, and 0.70 m in thickness. The perimetral turreted wall has a height above ground varying from 1.60 to 4.30 m, with an original estimated value from 4.00 to 5.30 m. The turreted wall has a thickness above the ground from 2.00 to 4.50 m. The slopes of the facings vary from 2.55° to 7.45°. The entrance stone lintel has an estimated height of approximately 2.40 m. The estimated area of the internal courtyards F, O, and P is 290 m^2. The maximum dimensions of the blocks at the base are 1.90 m width, 0.85 m height, and 1.40 m thickness, and the minimum dimensions are 0.70 m width, 0.40 m height, and 0.80 m thickness. The monument footprint area above the ground is S = 934 m^2. The estimated original footprint is S = 1040 m^2. The maximum dimensions of the above-ground structure are L = 41.10 m × l = 38.74 m × h = 9.70 m. The above-ground volume is V = 2735 m^3.

3.3. Accuracy Assessment

The accuracy of the MVS model can be estimated with the distribution and density of the SfM model tie points, which affects its MVS rebuild quality [17,19]. A uniform distribution of these points on all images contributes to more precise results in dense reconstruction with greater completeness and consistency in 3D reconstruction. The assessment of their density and distribution is an indicator of process quality, especially in the presence of regions with different types of surface textures and in the presence of areas with vegetation and occlusions. In general, the presence of vegetation of various types and densities in the relief area causes occlusion phenomena on the ground that result in a lower accuracy of the 3D model SfM-MVS. To obtain accurate 3D models from SfM-MVS data in vegetation environments, it is critical to consider and eventually remove areas where vegetation may obstruct the view of the underlying terrain. The AI classifier in RC allowed for the division of the entire 3D scene into categories such as "terrain", "monument elements", and "vegetation". In RC, it was possible to segment the areas with vegetation and then interpolate the mesh, reducing the effect of the occluded parts. Integrating nadiral and oblique shots from different angles can partially mitigate the occlusion problem and reduce approximations in 3D reconstruction. The assessment of the accuracy of the model was made on the surfaces of the monument without or with a limited presence of vegetation. Photogrammetric studies estimated errors in the determination of ground control points (GCPs) between 1 and 40 mm, with precisions from 1:300 to <1:1000 for objects with dimensions of $2 \leq D \leq 40$ m. Control measures were acquired using the total station (TS) [13]. Other studies have examined the accuracy of the 3D vertices of mesh derived from Multi-View Stereo (MVS) with precisions up to 1:300–1:1500 for larger areas with buildings in the range of $1 \leq D \leq 10$ m and errors in the range of $0.4 \leq Err \leq 18$ mm. Precisions ranged from 1:2500 to 1:5000 for small objects in the range of $1 \leq D \leq 0.2$ m, and errors were in the range of $0.02 \leq Err \leq 0.7$ mm [13,24,25]. Errors are reported as MEAN, STD, MAE, and RMSE. The use of encoded targets significantly improves the accuracy of the SfM-MVS 3D process by allowing error assessment both globally (reliability of measurements throughout the site) and locally (reliability of reconstructed surfaces from close-up photos) [13]. Using encoded targets, identified with "automatic detection" in RC in the SfM alignment phase, the calculated residue value on GCPs and CPs is lower than those obtained using the set of manually defined constraint points. RTK-GNSS measurements on both natural and coded targets have a similar mean residue variability range. The comparison between the data calculated in the SfM-MVS 3D process and the RTK-GNSS measurements on 9 GCPs and 18 CPs distributed throughout the site shows the residues and statistics in the X (east), Y (north), and Z (height) components. Georeferencing accuracy in SfM on GCPs was calculated with $RMSE_{XY}$ values of 16 mm and $RMSE_{XYZ}$ 19 mm. Instead, the position accuracy on the CPs was calculated in SFM with $RMSE_{XY}$ values of 20 mm and $RMSE_{XYZ}$ 25 mm. The values obtained are comparable with those obtained in other relevant applications and 3D modelling in the archaeological field. The positioning network of GCPs and CPs defined on the monument and in the surrounding area informed the calculation of quality performance and the minimization of the magnitude of errors within the processed point cloud. By analyzing the differences between 3D SfM-MVS coordinate values and RTK-GNSS measurements, the results highlight the quality of the horizontal and vertical precision achieved. The value of the RMSE precision is within ±20 mm for the RTK-GNSS measurements of the horizontal component and just over ±25 mm for the vertical component (Table 4). Therefore, 3D error is generally affected by higher values of this component. We have observed that the maximum elevation error is found in the highest area of the monument, while the lowest error is in the flat ones.

Table 4. MEAN, STD, MAE, RMSE, $RMSE_{XY}$, and $RMSE_{XYZ}$ error estimation calculated on the GCPs and CPs coordinates during the SfM-MVS process. For each point, the maximal projection error in RC is given in pixels.

GCPs	X (East) (m)	Y (North) (m)	Z (Height) (m)	Error (pix)	CPs	X (East) (m)	Y (North) (m)	Z (Height) (m)	Error (pix)
GCP0	0.000	0.003	0.004	0.227589	CP1	−0.003	0.001	0.004	0.301026
GCP2	−0.001	−0.003	0.000	0.204648	CP3	−0.010	0.035	−0.018	0.275347
GCP5	0.022	0.028	0.017	0.281916	CP4	0.022	0.021	0.022	0.457464
GCP8	−0.027	−0.008	0.018	0.784776	CP6	−0.004	−0.020	−0.023	0.756804
GCP11	−0.002	−0.001	−0.006	0.557855	CP7	−0.021	0.013	0.010	0.312578
GCP12	0.000	0.000	−0.001	0.545884	CP9	−0.011	−0.015	−0.018	0.369149
GCP13	0.002	0.000	0.001	0.347889	CP10	0.000	0.001	0.000	0.341045
GCP15	−0.001	−0.002	−0.002	0.547765	CP14	0.000	0.005	0.001	0.476852
GCP16	−0.002	−0.003	−0.004	0.497452	CP17	0.014	−0.018	0.005	0.388756
MEAN	−0.001	0.002	0.003		CP18	0.024	−0.003	0.023	0.658822
STD	0.012	0.010	0.009		CP19	0.011	0.023	−0.010	0.295589
MAE	0.007	0.006	0.006		CP20	−0.025	−0.008	0.015	0.775562
RMSE	0.012	0.010	0.009		CP21	−0.010	−0.013	−0.011	0.658825
$RMSE_{XY}$	0.016				CP22	0.012	−0.016	0.015	0.612578
$RMSE_{XYZ}$	0.019				CP23	−0.010	−0.013	−0.022	0.425996
					CP24	−0.001	−0.003	−0.003	0.368952
					CP25	0.020	0.002	0.004	0.796852
					CP26	−0.005	−0.004	0.007	0.774135
					MEAN	0.000	−0.001	−0.002	
					STD	0.014	0.015	0.014	
					MAE	0.011	0.012	0.012	
					RMSE	0.014	0.015	0.014	
					$RMSE_{XY}$	0.020			
					$RMSE_{XYZ}$	0.025			

The estimation of the shape errors in terms of the 3D model's rigidity was based on Euclidean distances calculated on the 3D Delaunay triangulation of GCPs (orientation) and CPs (validation) of known coordinates to evaluate the global geometric accuracy of the SfM-MVS 3D model (Figure 14). This method highlights any local deformations that can propagate on the dense point cloud. The calculated RMSE provides a statistical estimation of the error but is sensitive to the presence of systematic errors [17,35]. Low RMSE values indicate that the SfM-MVS 3D model is geometrically accurate and free of significant distortion or systematic errors (Table 5). In particular, the value of the RMSE for the Euclidean distances calculated on CPs was <20 mm with a maximum accuracy value of 1/4000. This value can be considered acceptable for this study compared to the extent of the surveyed area and the size of the monument.

The surface deviation errors estimation is based on a validation method of 3D SfM-MVS models without using the laser scanner as "ground truth". A photogrammetric benchmark was set as a reference model with the same APS-C sensor used for the survey. On the typical surface (consisting of the face of a tower of the perimetral wall), the photogrammetric clouds obtained using this APS-C sensor achieve results comparable to those of a full-frame DSLR camera with an absolute cloud-to-cloud distance (C2C) error of <1.44 cm at the 95th percentile (Q_{95}) after removing outlier values with errors higher than the 99th percentile (Q_{99}). These values are within the limits required for mapping architectural heritage. C2C comparison operating on the XY and XZ planes has the advantage of highlighting any rough errors present in the 3D data. Major errors on individual XY and XZ planes emerge immediately, affecting the overall statistical distribution of distances with larger values. Thus, with such a comparison, we could effectively detect both systematic differences in accuracy and significant localized errors in the 3D point clouds where there are occlusion areas with effects on the completeness of reconstructed surfaces.

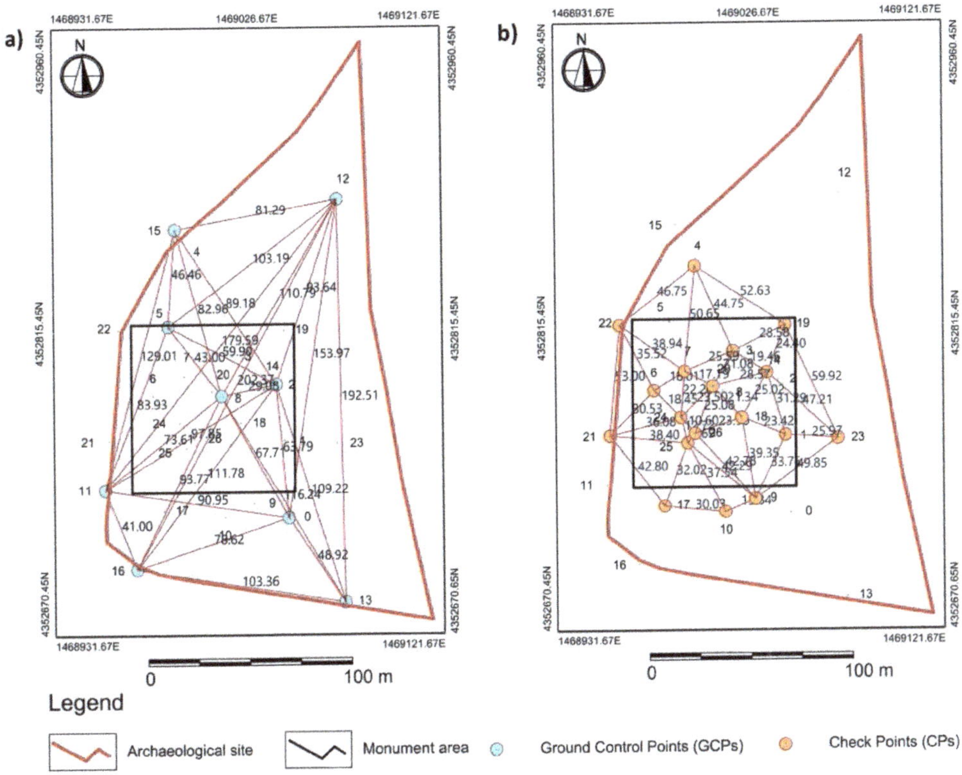

Figure 14. (**a**) View of the 3D Delaunay triangulation of GCPs (orientation) and (**b**) CPs (validation) to perform the error estimation of the SfM-MVS 3D model based on Euclidean distances.

Table 5. Estimation of RMSE errors based on Euclidean distances computed with SfM-MVS process and measured values (MD = measured distance, ΔD = absolute error difference).

GCPs (MD) (m)	ΔD (m)	CPs (MD) (m)	ΔD (m)	CPs (MD) (m)	ΔD (m)	CPs (MD) (m)	ΔD (m)
29.08	0.010	6.69	0.005	32.15	0.022	52.63	0.021
41.00	0.012	10.60	0.010	33.75	0.018	53.00	0.013
43.00	0.014	12.72	0.010	35.52	0.022	55.36	0.022
46.46	0.017	15.84	0.014	35.62	0.023	56.02	0.019
48.92	0.019	17.19	0.010	36.08	0.018	57.03	0.017
59.90	0.018	18.01	0.017	37.54	0.019	57.19	0.023
63.79	0.016	18.45	0.015	37.99	0.018	58.20	0.021
67.71	0.017	19.46	0.010	38.40	0.017	58.77	0.022
73.61	0.019	21.08	0.020	38.94	0.018	59.92	0.019
78.62	0.021	21.34	0.020	39.35	0.020	61.07	0.024
81.29	0.015	22.20	0.015	40.31	0.021	62.87	0.026
82.96	0.016	23.42	0.018	40.71	0.020	65.31	0.024
83.93	0.019	23.50	0.015	42.28	0.018	67.35	0.029
89.18	0.018	23.90	0.015	42.29	0.016	71.54	0.027
90.95	0.021	24.40	0.021	42.78	0.014	73.75	0.028
93.64	0.019	25.02	0.020	42.80	0.018	73.81	0.028

Table 5. *Cont.*

GCPs (MD) (m)	ΔD (m)	CPs (MD) (m)	ΔD (m)	CPs (MD) (m)	ΔD (m)	CPs (MD) (m)	ΔD (m)
93.77	0.022	25.08	0.010	43.22	0.019	81.21	0.030
97.85	0.020	25.59	0.017	43.65	0.017	90.93	0.032
103.19	0.019	25.97	0.015	44.75	0.017	100.81	0.030
103.36	0.021	28.57	0.016	46.75	0.021	111.83	0.035
109.22	0.024	28.58	0.018	46.87	0.019	-	-
110.79	0.023	29.35	0.022	47.21	0.018	-	-
111.78	0.023	29.89	0.018	47.22	0.019	-	-
116.24	0.024	30.02	0.019	47.24	0.017	-	-
129.01	0.023	30.03	0.022	48.03	0.020	-	-
153.97	0.022	30.53	0.015	49.85	0.018	-	-
179.59	0.017	30.86	0.018	50.65	0.024	-	-
192.51	0.022	31.29	0.017	51.38	0.027	-	-
202.37	0.025	32.02	0.019	52.42	0.019	-	-
RMSE	0.019					RMSE	0.020

Benchmarks on the typical surfaces were used for both SfM-MVS validation and the geometric simplification of the 3D model. Through the C2C comparison of the typical surfaces representing the geometric variability of the wall faces, the surface deviation error of the SfM-MVS model was estimated and the simplified process was guided with the combined use of RC and CC software (Figure 15). On the surface of tower M with a simplified edge length of 2 cm, the C2C error was <2.30 cm at Q_{95} after removing outlier values with errors higher than the Q_{99} percentile. With a simplified edge length of 3 cm, the C2C error was <2.9 cm at Q_{95} after removing outlier values with errors higher than Q_{99} (Tables 6 and 7). This integrated methodology allowed for the identification of the optimal level of simplification of the model (minimum length of the edges), keeping the error within the limits set by the study according to the specifications of the survey and the digital representation of the monument. In addition, the texture reprojection in RC from the original models to the simplified models allowed the high visual quality of the 3D reconstruction to be maintained.

Table 6. "Ground Truth" data acquisition setting and cloud-to-cloud (C2C) deviation analysis on M tower. Mean (M), standard deviation (STD), root mean square (RMS), and absolute errors on fitted Gaussian distribution at 95th percentile (Q_{95}).

Photos n	H (m)	GSD (cm/pix)	RR_4 (m)	Dw (m)	Dh (m)	B/H	Cloud-to-Cloud (C2C) (Fitted Gaussian Distribution)	M (m)	STD (m)	RMS (m)	Q_{95} (m)
35	5	0.13	8 × 5 × 2 m	6.48	4.29	0.2	XYZ	0.007	0.004	0.009	0.014

Table 7. Progressively simplified minimal edge length. "Ground Truth" and survey cloud-to-cloud (C2C) deviation analysis estimation. Mean (M), standard deviation (STD), root mean square (RMS), and absolute errors on fitted Gaussian distribution at 95th percentile (Q_{95}).

Simplify Minimal Edge Length (cm)	Cloud-to-Cloud (C2C) (Fitted Gaussian Distribution)	M (m)	STD (m)	RMS (m)	Q_{95} (m)
1	XYZ	0.007	0.004	0.009	0.014
2	XYZ	0.013	0.006	0.014	0.023
3	XYZ	0.018	0.007	0.019	0.029

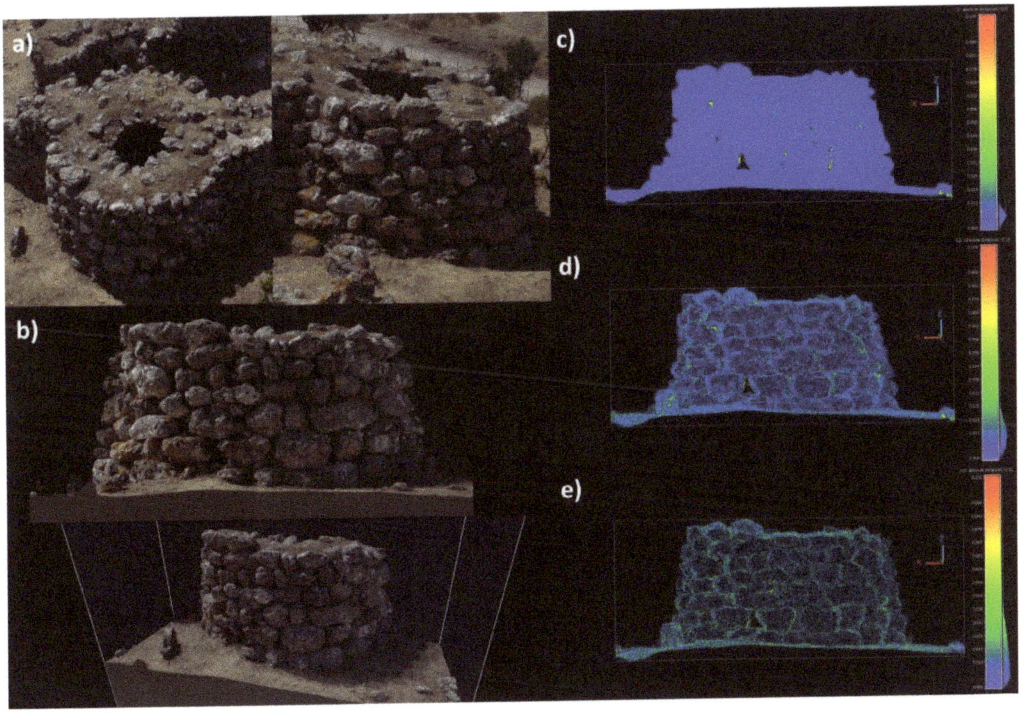

Figure 15. (**a**) Top view from north of detail tower M; (**b**) example of model surface for C2C comparison: the wall of tower M, with the 3D model delimited by the reconstruction region (RR_4); (**c**) the calculated distance from C2C of point clouds M1 and M2; (**d**,**e**) the calculated C2C error distance from point clouds with progressive simplification of the minimal edge length.

4. Discussion

Georeferencing was a key moment in the 3D reconstruction and topographic mapping of the surveyed area. The errors in the photogrammetric survey for GCPs and CPs are less than or equal to those of RTK-GNSS point acquisition. The workflow employed here allowed for the systematic definition of flight parameters, the photogrammetric acquisition of aerial and terrestrial schemes for the APS-C sensor used, the definition of the best distribution of GCPs, and the processing of photogrammetric data for the 3D modelling of the monument [13,18,27,28,31,33]. From the results obtained by comparing the compositional characteristics of the building materials with the geomaterials sampled during the geoarchaeological survey, it is evident that blocks from the slopes of Mount Acqua and the deposits of the Rio San Giovanni riverbed, worked by the action of the river current, were used for the construction of the Nuraghe. Many of these blocks, in fact, show a very close similarity in shape and size with those used in the Nuraghe. The definition of the typological characterization of the materials and the attribution of provenance areas are decisive in enriching the knowledge of the technological level reached by the Nuragic civilization. It is evident that although the stone materials used are not suited for the best shape regularization, due to their extreme hardness, the monument was built with an extremely advanced knowledge of building statics. The well-established construction technique of Nuraghes was masterfully adapted here to take advantage of the natural shapes of the boulders, without necessarily having to reshape them, enabling the construction of this impressive structure. Despite the widespread collapses and discontinuities, thanks to the processing of 2D and 3D digital models, the architecture of the monument is now well defined, allowing us to provide useful information to support future historical and archaeological studies regarding the actual proportions and dimensions that the monument must have had in its period of full functionality. The elaboration of the TIN of the basement allowed us to estimate the elevation of the rocky outcrops at the foundation of the Nuraghe and to define the lower levels of the entrances, which are currently also not visible due to the covering of materials from the numerous collapses of the structures. This also allowed us to estimate, with reasonable certainty, the thicknesses of collapse material that would have to be removed to obtain to the original floor level in the interior chambers. This also allowed us to obtain numerical data to propose a hypothesis about the actual original dimensions of the monument.

5. Conclusions

The proposed work represents part of a larger project on 3D reconstruction and virtualization in historical and archaeological settings, from the scale of the find to the scale of the site. This is to gain a better understanding of the state of preservation through dimensional analysis based on digital models and to hypothesize the original dimensions and shapes through the analysis of the archaeological remains. The approach used to perform the photogrammetric survey of the archaeological area in question and the workflow adopted in the photogrammetric process resulted in accurate 3D models according to the study specifications. In this study, we evaluated the accuracy and quality of the SfM-MVS 3D reconstruction of a Nuraghe using images acquired through aerial (UAV) and terrestrial surveys with an APS-C sensor. The validation process of the SfM-MVS model, through the estimation of position, rigidity, and surface deviation errors, allowed us to assess the accuracy and precision in terms of the robustness and completeness of the 3D model. Based on the study's objectives, the global geometric consistency of the reconstructed model and the visual quality of its surfaces were verified. We also demonstrated how the treatment and simplification of point clouds enabled the optimization of the obtained digital models. The results of this study show that for structures like those surveyed and under specific survey conditions, the use of high-quality APS-C sensors provides operational flexibility, allowing the acquisition configuration to be adapted to the specific needs of the survey. The versatility of APS-C sensors, which can be mounted on UAVs, telescopic masts, tripods, and three-axis stabilizers, enables efficient image capture under optimal operating conditions. This adaptability ensures the comprehensive coverage of the surveyed site in terms of the acquisition of images, improving the quality of the 3D reconstruction. The analysis of building materials and the geolithological context led to the identification of the areas of supply of local geomaterials used for the construction of the Nuragic monument. Of these, in particular, the blocks were skillfully selected for shape and size so that they could be used as is, without any reshaping. The results of our research provide the basic and indispensable documentation for the planning of maintenance and securing of the archaeological site, for the planning of future excavation campaigns, and for the planning of conservation restoration work. As of now, the 3D digital drawings have also been earmarked for purposes of enhancing the archaeological site through digital media and for promotion for tourism purposes. Future developments in our research will be oriented toward testing different data processing strategies with SfM-MVS integrated with LiDAR and laser scanner technologies. Different types of monuments and artefacts at different scales will be studied employing systems using new volumetric rendering techniques and advanced 3D visualization. The collection, management, and analysis of the photographic datasets according to the proposed workflow, in addition to representing a digital asset for each monument studied, will be able to improve, in a multidisciplinary context, the potential for the investigation and analysis of archaeological artefacts and Nuragic sites for the enhancement of cultural heritage.

Author Contributions: Conceptualization of photogrammetric methodology integrated with contribution to material analysis, S.C., C.M. and P.V.; photogrammetric survey planning and data acquisition, S.C. and C.M.; processing methodology, interpretation, and visualization of photogrammetric data, C.M.; XRD analysis, geoarchaeological survey, and sampling, S.C. All authors analyzed and discussed the results of the work and contributed to the drafting and revision of the manuscript. All authors have read and agreed to the published version of the manuscript.

Funding: The APC was funded by the Fondazione di Sardegna, grant number F73C22001270007 (P. Valera).

Data Availability Statement: Data are contained within the article.

Acknowledgments: We are very grateful to the Archaeological Superintendents of Cagliari and Oristano and to the Consorzio Natura Viva Sardegna of Domusnovas for providing access to the archaeological area. We also wish to thank Michal Jancosek for supporting our research with the advanced photogrammetry software RealityCapture.

Conflicts of Interest: The authors declare no conflicts of interest.

References

1. Lilliu, G. *I Nuraghi. Torri Preistoriche Della Sardegna*; Cambridge University Press: Cambridge, UK, 1962.
2. Lilliu, G. *La Civiltà dei Sardi dal Paleolitico alla Fine dell'età Nuragica*; Maestrale: Torino, Italy, 1988.
3. Contu, E. *L'architettura Nuragica*; 1981 in AA.VV; Libri Scheiwiller: Milano, Italy, 1981; pp. 1–178.
4. Contu, E. *La Sardegna Preistorica e Nuragica*; Carlo Delfino Editore: Sassari, Italy, 1998; Volume I–II.
5. Melis, P. I Nuraghi. In *la Sardegna Nuragica. Storia e Monumenti, Regione Autonoma della Sardegna/Carlo Delfino Editore*; Moravetti, A., Melis, P., Foddai, L., Alba, E., Eds.; Carlo Delfino Editore: Sassari, Italy, 2017; pp. 29–53, ISBN 978-88-7138-995-0.
6. Ugas, G. La Sardegna Nuragica. Aspetti Generali. In *LA SARDEGNA NURAGICA, Storia e Materiali*; Moravetti, A., Alba, E., Foddai, L., Eds.; Carlo Delfino Editore: Sassari, Italy, 2014; pp. 11–34, ISBN 978-88-7138-750-5.
7. Cossu, T.; Perra, M.; Usai, A. *Il Tempo dei Nuraghi. La Sardegna dal XVIII al VIII Secolo a.C.*; Ilisso: Cagliari, Italy, 2018; ISBN 978-88-6202-366-5.
8. Usai, A. *The Nuragic Adventure: Monuments, Settlements and Landscapes*; McDonald Institute for Archaeological Research: Cambridge, UK, 2021; pp. 155–169. [CrossRef]
9. Perra, M. I modelli di nuraghe come memoria collettiva di una civiltà. In *LA SARDEGNA NURAGICA, Storia e Monumenti*; Moravetti, A., Melis, P., Alba, E., Foddai, L., Eds.; Carlo Delfino Editore: Sassari, Italy, 2017; pp. 67–84, ISBN 978-88-7138-995-0.
10. Della Marmora, A. *Voyage En Sardaigne, Ou, Description Statistique, Physique Et Politique De Cette Île: Avec Des Recherches Sur Ses Productions Naturelles Et Ses Antiquités*; Biodiversity Heritage Library: Washington, DC, USA, 2022; Volume 1, ISBN 978-1017400939.
11. Della Marmora, A. *Itinerario dell'Isola di Sardegna (Titolo Originale "Itinéraire de l'Ile de Sardaigne, Pour Faire Suite au Voyage en Cette Contrée, Tome I–II)*; Ilisso: Cagliari, Italy, 1860; ISBN 9788862021081.
12. Vanzetti, A.; Castangia, G.D.; Depalmas, A.; Ialongo, N.; Leonelli, V.; Perra, M.; Usai, A. Complessi fortificati della Sardegna e delle isole del mediterraneo occidentale nella protostoria. In *Scienze Dell'antichità, Sapienza Università di Roma, Dipartimento di Scienze Dell'antichità*; Edizioni Quasar: Roma, Italy, 2013; Volume 19, pp. 2–3.
13. Sapirstein, P. Accurate measurement with photogrammetry at large sites. *J. Archaeol. Sci.* **2016**, *66*, 137–145. [CrossRef]
14. Kingsland, K. Comparative analysis of digital photogrammetry software for cultural heritage. *Digit. Appl. Archaeol. Cult. Heritage* **2020**, *18*, e00157. [CrossRef]
15. Massimiliano, P.; Domenica, C. UAV Photogrammetry and 3D Modelling of Complex Architecture for Maintenance Purposes: The Case Study of the Masonry Bridge on the Sele River, Italy. *Period. Polytech. Civ. Eng.* **2021**, 191–203. [CrossRef]
16. Marín-Buzón, C.; Pérez-Romero, A.; López-Castro, J.L.; Jerbania, I.B.; Manzano-Agugliaro, F. Photogrammetry as a New Scientific Tool in Archaeology: Worldwide Research Trends. *Sustainability* **2021**, *13*, 5319. [CrossRef]
17. Janiszewski, M.; Torkan, M.; Uotinen, L.; Rinne, M. Rapid Photogrammetry with a 360-Degree Camera for Tunnel Mapping. *Remote. Sens.* **2022**, *14*, 5494. [CrossRef]
18. Pepe, M.; Alfio, V.S.; Costantino, D. UAV Platforms and the SfM-MVS Approach in the 3D Surveys and Modelling: A Review in the Cultural Heritage Field. *Appl. Sci.* **2022**, *12*, 12886. [CrossRef]
19. Vlachos, M.; Berger, L.; Mathelier, R.; Agrafiotis, P.; Skarlatos, D. Software comparison for underwater archaeological photogrammetric applications. *Int. Arch. Photogramm. Remote Sens. Spat. Inf. Sci.* **2019**, *XLII-2/W15*, 1195–1201. [CrossRef]
20. Lingua, A.; Noardo, F.; Spanò, A.; Sanna, S.; Matrone, F. 3D model generation using oblique images acquired by UAV. *Int. Arch. Photogramm. Remote Sens. Spat. Inf. Sci.* **2017**, *XLII-4/W2*, 107–115. [CrossRef]
21. Cara, S.; Fiori, M.; Matzuzzi, C. Assessment of landscape by photogrammetry proximity UAV survey technique: A case study of an abandoned mine site in the Furtei area (Sardinia-Italy). In Proceedings of the 23rd International Mining Congress and Exhibition of Turkey, IMCET, Antalya, Turkey, 16–19 April 2013; Volume 1, pp. 83–92.

22. Cara, S.; Fiori, M.; Matzuzzi, C. Unmanned aerial vehicle system-based remote sensing for monitoring landscape degradation. In Proceedings of the International Multidisciplinary Scientific GeoConference Surveying Geology and Mining Ecology Management, SGEM 2017, Vienna, Austria, 27–29 November 2017; Volume 17, pp. 309–316.
23. James, M.R.; Robson, S. Mitigating systematic error in topographic models derived from UAV and ground-based image networks. *Earth Surf. Process. Landf.* **2014**, *39*, 1413–1420. [CrossRef]
24. Cara, S.; Valera, P.; Matzuzzi, C. Morphometric Analysis through 3D Modelling of Bronze Age Stone Moulds from Central Sardinia. *Minerals* **2021**, *11*, 1192. [CrossRef]
25. Casula, G.; Fais, S.; Cuccuru, F.; Bianchi, M.G.; Ligas, P. Diagnostic Process of an Ancient Colonnade Using 3D High-Resolution Models with Non-Invasive Multi Techniques. *Sensors* **2023**, *23*, 3098. [CrossRef] [PubMed]
26. Tahar, K.N. Height accuracy based on different RTK GPS method for ultra-light aircraft images. In Proceedings of theThe International Archives of the Photogrammetry, Remote Sensing and Spatial Information Sciences, Volume XL-3/W3, 2015 ISPRS Geospatial Week, La Grande Motte, France, 28 September–3 October 2015; pp. 287–292. [CrossRef]
27. Mora, O.E.; Suleiman, A.; Chen, J.; Pluta, D.; Okubo, M.H.; Josenhans, R. Comparing sUAS Photogrammetrically-Derived Point Clouds with GNSS Measurements and Terrestrial Laser Scanning for Topographic Mapping. *Drones* **2019**, *3*, 64. [CrossRef]
28. Barba, S.; Barbarella, M.; Di Benedetto, A.; Fiani, M.; Gujski, L.; Limongiello, L. Accuracy Assessment of 3D Photogrammetric Models from an Unmanned Aerial Vehicle. *Drones* **2019**, *3*, 79. [CrossRef]
29. Lai, L.; Sordini, M.; Campana, S.; Usai, L.; Condò, F. 4D recording and analysis: The case study of Nuraghe Oes (Giave, Sardinia). *Digit. Appl. Archaeol. Cult. Herit.* **2015**, *2*, 233–239. [CrossRef]
30. Moyano, J.; Nieto-Julián, J.E.; Bienvenido-Huertas, D.; Marín-García, D. Validation of Close-Range Photogrammetry for Architectural and Archaeological Heritage: Analysis of Point Density and 3D Mesh Geometry. *Remote Sens.* **2020**, *12*, 3571. [CrossRef]
31. RealityCapture Software, Version 1.3.1.117316 [CLI Software]. Epic Game Slovakia s.r.o., Slovakia. Available online: https://www.capturingreality.com (accessed on 20 March 2024).
32. Ezequiel, F.G.; Aguera-Vega, F.; Carvajal-Ramirez, F.; Martinez-Carricondo, P. UAV Photogrammetry Accuracy Assessment for Corridor Mapping Based on the Number and Distribution of Ground Control Points. *Remote Sens.* **2020**, *12*, 2447. [CrossRef]
33. Cucci, D.A. Accurate optical target pose determination for applications in aerial photogrammetry. In Proceedings of the SPRS Annals of the Photogrammetry, Remote Sensing and Spatial Information Sciences, XXIII ISPRS Congress, Prague, Czech Republic, 12–19 July 2016; Volume III-3; pp. 257–262. [CrossRef]
34. Guo, X.; Chen, Y.; Wang, C.; Cheng, M.; Wen, C.; Yu, J. Automatic shape-based target extraction for close-range photogrammetry. In Proceedings of the International Archives of the Photogrammetry, Remote Sensing and Spatial Information Sciences, Volume XLI-B1, 2016 XXIII ISPRS Congress, Prague, Czech Republic, 12–19 July 2016; pp. 583–587. [CrossRef]
35. Nocerino, E.; Stathopoulou, E.K.; Rigon, S.; Remondino, F. Surface Reconstruction Assessment in Photogrammetric Applications. *Sensors* **2020**, *20*, 5863. [CrossRef] [PubMed]
36. MikroKopter, 2024 HiSystems GmbH. Website for MikroKopter Do-It-Yourself Construction. Available online: https://doc.mikrokopter.de/ (accessed on 20 March 2024).
37. Wojciech, A.D. Exploiting the Redundancy of Multiple Overlapping Aerial Images for Dense Image Matching Based Digital Surface Model Generation. *Remote Sens.* **2017**, *9*, 490. [CrossRef]
38. Jancosek, M.; Shekhovtsov, A.; Pajdla, T. Scalable Multi-View Stereo. In Proceedings of the IEEE 12th International Conference on Computer Vision, ICCV Workshops, Kyoto, Japan, 27 September–4 October 2009; pp. 1526–1533. [CrossRef]
39. Jancosek, M.; Pajdla, T. Multi-view reconstruction preserving weakly-supported surface. In Proceedings of the IEEE Conference on Computer Vision and Pattern Recognition, Colorado Springs, CO, USA, 20–25 June 2011. [CrossRef]
40. Cloud Compare, Version 2.12.4 [GNU GPL Software]. 2021. Available online: http://www.cloudcompare.org (accessed on 20 March 2024).
41. Note Illustrative della Carta Geologica d'Italia 1:50.000, Foglio 555 Iglesias, 2015, ISPRA, Servizio Geologico d'Italia, Regione Autonoma della Sardegna. Available online: https://hdl.handle.net/20.500.11769/103033 (accessed on 20 March 2024).

minerals

MDPI

Article

An Integrated Petrographic, Geomatic and Geophysical Approach for the Characterization of the Carbonate Rocks of the Calcari di Cagliari Formation

Giuseppe Casula [1,*], Silvana Fais [2,3,4], Francesco Cuccuru [2], Maria Giovanna Bianchi [1] and Paola Ligas [2]

1 Istituto Nazionale di Geofisica e Vulcanologia (INGV)—Sezione di Bologna, Viale Berti Pichat 6/2, 40127 Bologna, Italy; mariagiovanna.bianchi@ingv.it
2 Department of Civil, Environmental Engineering and Architecture (DICAAR), University of Cagliari, Via Marengo 2, 09123 Cagliari, Italy; sfais@unica.it (S.F.); cuccuru.f@gmail.com (F.C.); pligas@unica.it (P.L.)
3 Consorzio Interuniversitario Nazionale per l'Ingegneria delle Georisorse, CINIGEO, Palazzo Baleani, Corso Vittorio Emanuele II 244, 00186 Roma, Italy
4 National Research Council of Italy (CNR)—Institute of Environmental Geology and Geoengineering (IGAG), Via Marengo 2, 09123 Cagliari, Italy
* Correspondence: giuseppe.casula@ingv.it; Tel.: +39-0514151415

Abstract: Non-invasive techniques, such as close-range photogrammetry (CRP) and 3D ultrasonic tomography complemented with optical and scanning electron microscopy and mercury porosimetry, were applied to characterize the carbonate rock samples of the Calcari di Cagliari formation. The integrated approach started with the computation of high-resolution 3D models of the carbonate samples using the CRP technique to produce 3D high-resolution models texturized both with natural colors and intensity. Starting from the 3D models from previous techniques, a 3D ultrasonic tomography on each rock sample was accurately planned and carried out in order to detect the elastic properties of such rocks and relate them to textural heterogeneity or internal defects. The results indicate that the relationship between longitudinal velocity and rock properties is complex even in the same carbonate formation. Understanding the relationship between the geomatic and geophysical responses in the investigated rock properties, such as textural characteristics and especially structure and geometry of pores, type of pores, tortuosity and cementing material, is important for many practical applications and especially in the diagnostic process of the conservation state of monumental structures. The integration of the above non-invasive techniques complemented by petrographical–petrophysical data proved to be a powerful method to associate each lithotype with a different susceptibility to degradation. The results presented in this paper demonstrate that the proposed integrated use of complementary methodologies would guarantee the reproducibility of the measurements both at the laboratory and field scale for the monitoring in time of the rock condition while giving a useful contribution in making decisions on an appropriate remedial strategy.

Keywords: carbonate rocks; petrographic data; photogrammetry; ultrasonic tomography; integrated interpretation

1. Introduction

This paper focuses on the integrated application of geomatic and geophysical non-invasive techniques integrated with petrographical data for the experimental characterization of significant samples of carbonate rocks known as the Calcari di Cagliari formation. The integration of all data aimed to provide a better understanding of the possibility of disintegration and alteration involving the Calcari di Cagliari, and ultimately for the prevention and conservation of buildings of historical and artistic interest. In fact, most of the buildings in the historic center of the town of Cagliari were built using the above carbonate materials, which are comparable to other carbonate building materials used in the Mediterranean Basin in both construction and ornamentation [1–6].

Traditionally the characterization of carbonate materials has been carried out using optical microscopy (OM) and scanning electron microscopy (SEM) to analyze thin sections appropriately prepared from specimens [7–11]. Though indispensable for the important information provided, this approach is clearly destructive and there is a limit to the number of samples that can be lost. Furthermore, using traditional petrographic techniques alone, we cannot analyze the evolving processes of degradation by observing the shallow and in-depth conditions of the stone materials in time. In recent years, the use of non-invasive techniques has been developed in many fields of applied research [12–15] and especially in the analysis of stone materials [2,16–22]. Furthermore, non-invasive techniques are the most appropriate for assessing the quality of stone building materials in monumental and historic structures. For a successful on-site analysis, it is crucial to carry out a non-invasive characterization of samples from the same quarries of origin as the construction materials.

As is known, the non-invasive analysis of rock samples is becoming ever more widespread with the introduction of high-tech methods, such as X-ray computed tomography (X-ray CT), X-ray micro-computed tomography (X-ray micro-CT), digital radiography, micro X-ray fluorescence (μ-XRF) [23–25] and others. X-ray CT is a very effective non-destructive method for providing the three-dimensional structures of rock core samples [26] or complex artworks [27,28]. X-ray micro-CT and digital radiography are non-destructive techniques that provide high-resolution images of the internal structures of the investigated stone materials. In addition, micro-CT images can be elaborated for the realization of high-resolution 3D digital reconstructions and models. These high-tech methods provide a great amount of microstructural information. Furthermore, these high-tech methods are extensively used in hydrocarbon research to analyze the porous structure of rock samples, and in scientific drilling to help the interpretation of lithology and structure while preserving the digital record of the core prior to sacrificing it for subsequent destructive analyses [29–31]. Some degradation areas inside the materials can be also detected by means of neutron diffraction [32] and neutron tomography [33,34]. However, the above recent techniques involve very expensive equipment available in few technologically advanced laboratories.

In this study we have chosen to apply an integrated petrographic, geomatic and geophysical approach for stone material characterization to provide accurate, fast and reliable results at a resolution suitable for the preset scientific targets, but at the same time more accessible and at low cost, thus making it available for technicians working in the cultural heritage field.

Nowadays, among the geomatic methodologies used to compute high spatial resolution 3D models of objects of complex shapes, close-range photogrammetry (CRP) based on structure from motion (SfM) and multi view stereo matching (MVS) algorithms derived from computer vision are actually some of the most used methods in different fields of geomatics like terrain surface modeling, digital elevation modeling (DEM), digital terrain modeling (DTM), digital surface model (DSM) computations, architectural, historical and modern building information modeling (BIM) and so on [35,36]. In fact, SfM-based techniques have already proven to be successful for accurate 3D modeling of built structures, bare geological substrates and fine spatial scale individual plant structures [37,38]. This methodology can produce 3D models with properties similar to those of laser imaging detection and ranging (LIDAR), but with red–green–blue (RGB) spectral attributes for each point of the surface of the analyzed targets [39]. The technique consists essentially of the computation of high-resolution point clouds and corresponding meshes texturized with natural colors based on the inherent motion of numerous overlapping images taken from different locations, in particular arrays of good quality 2D images like photographs acquired with optical cameras in the terrestrial modality as well as in the aerial one with the aid of low-cost unmanned aerial vehicles (UAV). The terrestrial modality is used when high-resolution and performant optical cameras that are too heavy and bulky to be mounted on drones must be used; conversely, aerial surveys are used when parts of a studied complex-shaped object, like building roofs or areas of terrains of difficult access, are unreachable from the ground [39]. In this work, we used a low-cost commercial grade digital camera to compute high spatial resolution 3D models of significative samples of carbonate facies of the Calcari di Cagliari.

The above geomatic technique is very useful to analyze shallow parts of the stone materials and holds vast potential for improving both the quality and efficiency of a 3D digital model useful for the subsequent geophysical analyses. In fact, the high-resolution (HR) 3D models of the investigated carbonate samples have provided accurate information on their geometrical shape, for example, the presence of defects, small cavities of different origins, fractures and alteration zones in the shallow part of the investigated materials.

Geophysical techniques, such as the ultrasonic tomography applied in this study, can detect the elasto-mechanical conditions inside the materials while analyzing the propagation of elastic waves within them. The knowledge of the elastic properties of the carbonate rocks is quite challenging due to their heterogeneity and complexity.

Many researchers [18,22,40–45] have studied the relations between the longitudinal and/or transverse velocities with petrographic and petrophysical characteristics of the carbonate materials, using laboratory techniques based on the ultrasonic velocity test. An accurate petrographic study using optical and electronic microscopy can improve the interpretation of the non-invasive diagnostic techniques, and especially the deep understanding of the behavior of the ultrasonic signal propagation [9,46–51].

The advantage of the ultrasonic method is that it is easy to use, less costly and non-destructive. Furthermore, it can be used both in the laboratory and in the field and proves useful in many sectors of the applied research. In this study, we carried out a laboratory ultrasonic survey at a resonance frequency of 82 kHz by acquiring and processing 3D ultrasonic travel-time tomography data (namely P-wave travel times) on the significant samples of the Calcari di Cagliari formation. Three-dimensional ultrasonic tomography (3D UT) represents a high-resolution technique to inspect the inner parts of the investigated rock samples providing information on their elastic properties variations related to their petrophysical and mechanical properties. In planning and interpreting the 3D UT, the above geomatic techniques play an extremely important role, since they provide high-resolution 3D digital models to precisely locate a large quantity of measurements (e.g., locations of the ultrasonic signal transmission and receiver points). Furthermore, the 3D digital model obtained by SfM photogrammetry also represents an effective tool for the accurate and effective representation of the 3D tomographic results, allowing the precise locations of longitudinal velocity changes in relation to the textural variation of the materials and the presence of defects or altered zones inside them.

Taking into account the aspects considered above, the aim of this paper was to provide a framework for the applicability of some non-destructive techniques (NDT) supported by OM and SEM observations of thin sections cut from the same samples used for the non-destructive characterization. The OM and SEM have also been complemented by mercury intrusion porosimetry (MIP) to obtain further information on the structure of the porosity system of the investigated carbonate rocks.

2. Materials and Methods

2.1. Materials

The carbonate succession known as Calcari di Cagliari outcrops in the hills and along the coast of the city of Cagliari [52,53].

The late Miocene transgressive succession of the Calcari di Cagliari is characterized by a basal part in which Tortonian carbonate sediments (mudstones and wackestones) of an external platform have been deposited. With the progressive decreasing bathymetry that occurred in the upper Tortonian–lower Messinian, biogenic carbonate platforms of the littoral environment developed. The change in the sedimentation environment due to bathymetric variations led to three distinct carbonate facies, named from bottom to top, Pietra Cantone, Tramezzario and Pietra Forte. These were the names given to these lithotypes by the old quarrymen.

The Pietra Cantone is a yellow, poorly stratified, bioclastic limestone with a mud-supported texture due to a carbonate sedimentation that occurred at a paleobathymetry of about 60–80 m, belonging to the circalittoral plane [54]. This rock is characterized by a high content of planktonic and benthic foraminifera, and small amounts of iron oxides and terrigenous components such as quartz, feldspars and biotite. The fossiliferous content has allowed us to date this lithotype to the Tortonian age [52,55].

The Tramezzario is a white bioclastic limestone with a grain-supported (grainstone-packstone) texture, characterized by a textural heterogeneity with both reef facies rich in bioclasts and deeper sea facies rich in carbonate mud. However, the biocenoses that characterize the Tramezzario indicate a paleobathymetry of sedimentation not exceeding 40 m, belonging to the infralittoral–circalittoral bathymetric plane [54]. Isotope analyses performed with the 86Sr/87Sr method have provided an age referable to the end of the Tortonian [56].

The Miocene carbonate succession of the Calcari di Cagliari ends with the Pietra Forte, a cream-colored biohermal or biostromal massive limestone, typical of a littoral–infralittoral depositional environment with high energy and a paleobathymetry less than 30 m [54]. The deposition of the Pietra Forte is mainly due to the growth of Lithothamnium algae in which bioclasts of a different nature were incorporated, mainly bivalves, gastropods, crustaceans and colonies of bryozoans. Based on the faunas and the stratigraphic position, the Pietra Forte can be dated back to the Tortonian, but due to some affinity with similar formations in the Gulf of Oristano (Sardinia, Italy) it can also be referred to the Messinian [54,55].

The different intrinsic characteristics of each limestone facies affect the forms and intensity of degradation. Pietra Cantone is a poorly cemented rock, characterized by a microporous carbonate matrix which makes it hygroscopic and therefore susceptible to aggressive forms of decay such as alveolization (Figure 1a). The outcrops and artefacts made up of this lithotype are more susceptible than the other two lithotypes of the Calcari di Cagliari formation to degradation due to salt-rich aerosols from the sea. The saline solution permeates deeply into the rock and when the salts crystallize, they cause micro-fractures and a loss of material in the form of dust. The Tramezzario is usually a hard, compact rock, although in several cases it is characterized by the combination of weak cementation and high secondary porosity that can cause severe degradation forms. In particular conditions, such as high moisture levels, degradation occurs more intensely, causing detachment and the complete disintegration of the rock into carbonate mud (powder) (Figure 1b).

Figure 1. Degradation forms of the Calcari di Cagliari limestones: (**a**) alveolization of the Pietra Cantone; (**b**) loss of material and pulverization in a wall built with Tramezzario ashlars; (**c**) detachment and oxidation in a Pietra Forte ashlar.

The Pietra Forte is generally a very compact, tenacious and hard limestone. Due to its excellent cementation, it is less susceptible to decay. However, fractures, detachments and oxidation could affect this lithotype (Figure 1c). Fractures are often caused by stress exceeding the rock's strength that causes a loss of cohesion along the weakest planes of the rock. The detachment of material is generally due to the release of inner tensions after the rock is quarried and reduced to ashlars for building. Especially within the fractures, the Pietra Forte can present impurities containing iron minerals which, by oxidation, create red-brown patinas that render the altered material brittle and favor its detachment. The Pietra Forte is also affected by dissolution phenomena that produce macropores and superficial and hypogenic karst forms. The karst forms can be large and can extend over wide areas, making the Pietra Forte the main aquifer of the city of Cagliari.

2.2. Optical and Scanning Electron Microscopy

Thin sections of the three facies of Calcari di Cagliari were analyzed by optical microscopy (OM) and scanning electron microscopy (SEM) in order to identify their textural characteristics and the amount and type of pores, in particular micropores by SEM. OM analysis was carried out by means of a petrographic microscope in polarized light (Carl Zeiss Axioplan microscope—Carl Zeiss, Oberkochen, Germany). SEM was carried out by a Zeiss EVO 50 VP model scanning electron microscope (Carl Zeiss, Oberkochen, Germany) connected to an EDS X-Max (Oxford Analytical Instruments Ltd., High Wycombe, UK). SEM was performed under a high vacuum with an electron high tension (EHT) of 20 kV, a working distance (WD) of 10 mm, an Iprobe current of 200 pA, and a resolution of about 1–2 nm. The thin sections studied with OM were treated with blue dye epoxy resin to better visualize the pore space [22,48,57,58] and their connection at the meso-scale [59], while for SEM, the thin sections were metal-coated, creating a conductive 10 nm layer of gold on their surfaces to inhibit charging, reduce thermal damage and improve the secondary electron signal required for the morphology examination.

2.3. Mercury Intrusion Porosimetry

The knowledge on the textural features of the Calcari di Cagliari by OM and SEM was implemented with quantitative information on the pore structures obtained by the mercury intrusion porosimetry (MIP) technique. Fifteen cubic representative samples (with the largest dimension of 1.5 cm per side), five from each carbonate lithotype (Pietra Cantone, Tramezzario and Pietra Forte), were prepared and dried in an oven at 60 °C for 48 h for the MIP analyses. A Micromeritics Autopore IV 9500 (Micromeritics Instrument Corporation, Norcross, GA, USA) was used to measure the connected porosity for pore-throat diameters that cover the 3 nm–360 μm range, applying an initial pressure of 0.5 psia and reaching a maximum pressure of 60,000 psia to measure the smallest pore-throat diameters. Owing to MIP analyses, it was possible to quantify the number of connected pores (effective porosity), pore-throat size distributions (radii/diameters) and bulk and skeletal densities. Besides this information, some key petrophysical parameters of the pore network, such as permeability and tortuosity [60–62], were deduced. These parameters depend on the size and arrangement of pores/throats and pore-to-pore connections. Mercury intrusion porosimeter instruments can compute permeability and tortuosity using their specialist software by applying the well-known relationship proposed by Katz and Thompson [59,60], who introduced the K-T method to calculate permeability, and by Hager [61], who proposed a relationship for the calculation of tortuosity.

2.4. CRP Method

Data Acquisition—Three prismatic-shaped significative samples of each investigated lithotype, namely, Pietra Cantone, Tramezzario and Pietra Forte, of the Calcari di Cagliari formation were prepared with an average volume of about 20 × 10 × 10 cm^3. Two terrestrial photogrammetric campaigns were performed on two different days in this study: 30 December 2022 and 2 January 2023. A single lens commercial digital camera model, namely the Canon Power Shot 730 HS CMOS with a resolution of 20.3 Mpx, was operated to acquire sets of about a hundred high-quality 2D photo images for every rock sample, all around the samples under study, at short and predetermined distances (20–50 cm) and at different station points evenly spaced at a mean distance of 10 cm between them (about 80% overlapping). Finally, all images were collected under light conditions as homogeneous as possible and after positioning the studied sample on a calibrated base equipped with reference markers characterized by high reflectivity contrast and known dimensions [9,18,42].

Data Processing—At the pre-processing stage, the 2D images were first manually trimmed to remove unusable photos; they were then pre-processed with release 2.0.2 of the AgisoftMetashape® software (Agisoft LLC, St. Petersburg, Russia) based on structure from motion (SfM) methodology to avoid the use of noisy data, to finely align poses and to generate a highly dense point cloud.

Image alignment and sparse cloud generation—At the image orientation stage, the position of the camera is detected, the photo orientation is determined using both intrinsic camera parameters, such as focal length, coordinates of the image principal point and lens distortion coefficients, and extrinsic orientation parameters, such as the three translation components and the three Euler rotation angles estimated during the image orientation process. After the image orientation procedure is successfully completed, an a priori model in the form of a discrete point cloud is generated using essentially the reference points [63].

Dense cloud generation, meshing and texturing—At the following stage of dense cloud generation, a multi view stereo matching (MVS)-styled process is started, by which depth maps are calculated to perform the densification of points. The depth maps are computed essentially using the image pairs that overlap and taking into account the intrinsic and extrinsic orientation parameters derived from the bundle adjustment operations.

A combined depth map is generated, stacking pairwise depth maps computed for each camera that are subsequently used to generate partial dense point clouds. These clouds are then used to compute the final dense cloud. During this process, additional well-suited noise filter algorithms are applied step-by-step to the overlapping regions [39].

At this stage of the processing, planes are routinely fitted to groups of neighboring pixels of the combined depth maps in order to compute the normal of the partial dense clouds; finally, clouds can be texturized using the colors extracted from the images.

The processing is followed by the meshing operation that consists essentially of a 3D surface reconstruction process for quickly transforming the dense point cloud into a triangulated surface. Moreover, the necessary triangulated model optimization and simplification is performed by means of the application algorithms well-implemented in the AgisoftMetashape® software package [63].

An option to build the texture clouds is available in the recent versions of this package, by which the natural color contents of the photographic images can be used to texturize the 3D final models using an RGB color scale.

The three colorized, high density, registered point clouds computed using the procedure described above and representing in detail the rock samples under study were format converted into the e57 format (namely, the file format for 3D laser scanner data exchange) and then processed with the JRC 3D Reconstructor® Vs 4.1.2 (Gexcel, Brescia, Italy) and the CloudCompare Vs 2.12.4 free software. Clouds were also manually edited and then filtered to remove noisy data; the resulting unified clouds were then converted into mesh models texturized with both natural colors and light intensity to be useful as input for professional engineering-style software such as CAD® [16,49].

2.5. *Three-Dimensional Ultrasonic Tomography*

Based on the 3D models obtained with the above described geomatic technique (CRP-SfM), 3D ultrasonic tomography (3D UT) was planned on significative samples of each investigated lithotype (Pietra Cantone, Tramezzario and Pietra Forte).

The 3D UT adequately describes the effects of rock properties on the ultrasonic signal propagation and highlights the presence of potential defects such as cracks, fissures and cavities inside the materials. The morphology of the shallow materials in each sample was determined from the 3D high-resolution digital photogrammetric models in order to design an optimal ultrasonic survey and provide a very good spatial coverage of the investigated samples.

The carbonate samples were prepared in a prismatic shape with an average size of 20 cm × 10 cm × 10 cm. Ultrasonic measurements were carried out by the transmission method according to the ISRM 2007, 2014 [64,65] using a portable Ultrasonic

Non-Destructive Digital Indicating Tester (PUNDIT Lab plus) device (Proceq, Schwerzenbach, Switzerland) that measures the time of propagation of ultrasonic pulses within the 0.1–9999.9 μs range with a precision of 0.1 μs. The PUNDIT Lab plus was interfaced with the portable oscilloscope Fluke 96B for the acquisition of the digital signals emitted and received from two transducers at a resonance frequency of 82 kHz. The silicone snug sheets were used as the best coupling agent [66] to guarantee the best contact between the transducers and materials. The transducers were placed on all faces of each sample with 25 mm of vertical spacing along parallel vertical profiles to surround the entire sample. Each station (on average 72 stations per sample) was alternately used as a transmitter and receiver. The transit time of the longitudinal ultrasonic waves between stations (source–receiver path) was measured along a very large number of paths (approximately 2400) and averaged over eight determinations. The mean of the measurements was computed not considering values with a deviation higher than 3%. In this study, only the first arrivals for each source–receiver path were considered and confidently associated to the longitudinal wave (P-wave). The use of the ultrasonic longitudinal signal (P-wave) in velocity detection is recommended in practical applications, since the first arrival of the ultrasonic signal can be recognized unambiguously as a P-wave arrival. Generally, the quality of first-arrivals is related to many experimental conditions and especially to the signal-to-noise (S/N) ratio. In a few cases in which a low S/N ratio was found, a filtering operation was applied to improve it. We preferred to hand-pick the first arrivals of the longitudinal waves for the tomographic analysis. In fact, the accurate determination of the first-arrival times is needed to guarantee the high quality of the tomographic solutions. In many experimental conditions, the automated picking of first-arrivals can be a very difficult task especially in the case of a low S/N ratio.

The propagation velocity of the P-wave is detected from the ratio of the length of the ray-path (source–receiver distance) to the transit time of the longitudinal ultrasonic signal.

The 3D UT aimed to obtain a detailed elastic characterization of the carbonate samples consisted of the 3D reconstruction of the longitudinal velocity distribution inside the samples based on the P-wave travel times between the different source–receiver pairs precisely located on the base of the 3D high-resolution digital photogrammetric models. With many intersecting ray paths between different source and receiver positions around the sample, it is possible to reconstruct a detailed 3D field velocity of the elastic waves. In this study a tomographic inversion was applied to the ultrasonic data (measured travel times) to produce the 3D velocity model. This was determined starting from the P-wave first-arrival times as input data for the simultaneous iterative reconstruction technique (SIRT) [67,68], considering curved ray tracing to account for the heterogeneous velocity structure. The iterative reconstruction technique involves successive approximations to correct for an arbitrary initial velocity model. In order to obtain a realistic and not an arbitrary starting velocity model as input for the SIRT to invert travel-time data, a methodology based on the cross-correlation function (CCF) was used [69]. Starting from the initial velocity model, the SIRT algorithm is followed by computing the travel times for each source/receiver path. After this, the calculated and picked times are compared and a correction factor for the time discrepancy is applied to the velocity value of every cell (voxel) affecting the model. In this way, the initial velocity model is modified by repeated cycles, and the iterative process goes on until the necessary accuracy is achieved. In this study the entire processing sequence was repeated for 20 iterations which allowed it to obtain the final velocity model. This number of iterations was found appropriate to achieve the best representation of the 3D distribution of the longitudinal velocity within the investigated volumes and can be considered a good compromise between the resolving power and the reliability of the final velocity model.

The 3D rendering of the resulting velocity distribution inside the samples as a result of the tomographic inversion procedure was carried out using the Voxler v. 4.3.771 by Golden Software, which allows us to manipulate and organize large quantities of data in a convenient and effective manner. The 3D representation of the field velocity and the possibility to visualize it through the 3D data volume facilitate the analysis and interpretation of the results. The 3D volume visualization with zooming and rotation allows a better understanding of the relationship between velocity variations and material properties for analysis and accurate interpretation. Furthermore, it is possible to create in an interactive manner as many slices as necessary, vertical or horizontal, that also help the interpretation. By observing the development of the velocity variations from slice to slice, the interpreter can acquire a good feel of their spatial development in the investigated volume.

3. Results and Discussion

3.1. Thin Section Analysis

The textural characteristics of the studied lithotypes were analyzed by optical microscopy, with reference to the presence of matrix or cement, as well as content, type and arrangement of the allochems. OM observations were integrated and enhanced by SEM in order to define the bioclast packing and the different grains–matrix or grains–cement relationships, while pointing out the different types of primary and secondary pores. The porosity detected by OM can be classified as mesoporosity [70], while the pores analyzed by SEM can be classified as microporosity [70].

Pietra Cantone is a bioclastic limestone with a mud-supported or micritic texture, mainly characterized by the remnants of foraminifera (Figure 2a) and other bioclasts ranging from 100 μm to 2.5 mm. The matrix is a carbonate mud made up of microcrystals of micrite. Within the micrite cement, the presence of iron oxides and terrigenous components such as quartz, feldspar and biotite is common (Figure 2b). The presence of iron oxides gives this lithotype its typical yellow color. According to the carbonate rocks classification of Dunham (1962) [71], the Pietra Cantone can be classified as mudstone–wackestone. The mesoporosity values of Pietra Cantone observed with OM has values from 7% to 25%. The mesoporosity is characterized by primary interparticle pores that can evolve into forms of secondary porosity, such as vug porosity, due to dissolution processes (Figure 2c,d). The SEM observations showed that Pietra Cantone is characterized mainly by a micropore network of well-connected interparticle pores with a mean size of 2 μm (Figure 2e,f), concentrated in the micrite [18]. This primary microporosity in the mud matrix is the main cause that makes this rock hygroscopic and highly susceptible to degradation.

Tramezzario is a grain-supported limestone rich in bioclasts with sizes ranging from 500 μm to 4 mm. The bioclasts are mostly made from well-sorted, sub-rounded algal fragments (Figure 3a) and bioclasts of other organism such as mollusks, bryozoa and crustacea, supported by a sparry calcite cement (Figure 3b). Tramezzario, according to the Dunham classification (1962) [71], can be defined as grainstone–packstone. In Tramezzario, the porosity is 10% on average, but it can reach 20% in deep sea facies, richer in carbonate mud. The mesoporosity observed at OM is a secondary porosity mainly of the channel, moldic, intraparticle and fracture types (Figure 3c,d). Except for the fracture type, these pores occur due to dissolution processes and are well connected. With SEM, the Tramezzario is characterized by single subhedral–anhedral calcite microcrystals (sizes 2–3 μm) or by polyhedral aggregates of calcite. Intercrystal pores [72] with sizes between 4 μm and 5 μm (Figure 3e) represent the main microporosity of the Tramezzario. These pores, located within the sparry calcite cement are generally interconnected, representing the preferential path on which the phenomena of dissolution act. The dissolution, therefore, determines the development of secondary pores such as vugs (Figure 3f) and channels.

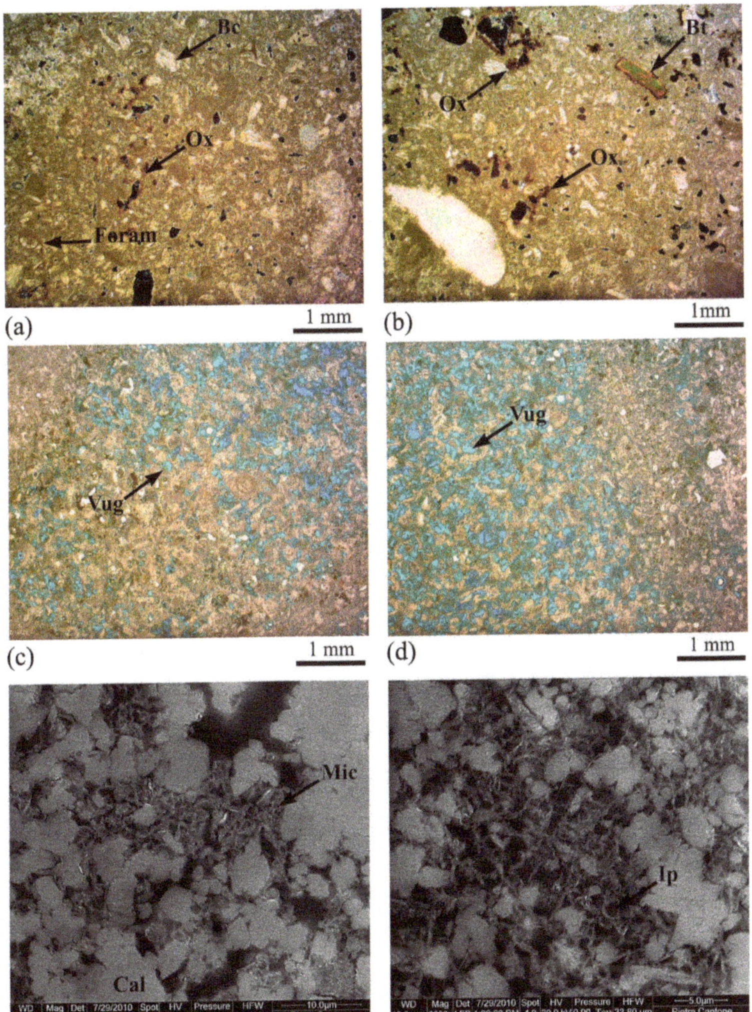

Figure 2. Textural characteristics of Pietra Cantone: (**a**) Mud-supported texture with bioclasts (Bc), foraminifera (Foram) and iron oxide (Ox); OM cross polarized light. (**b**) Iron oxide (Ox) and biotite (Bt) crystal; OM cross polarized light. (**c**) Vug secondary porosity (Vug); OM plane polarized light. (**d**) Vug secondary porosity (Vug) in a particularly porous sector of the rock; OM plane polarized light. (**e**) Micrite (Mic) microcrystals and calcite (Cal); SEM image. (**f**) Interparticle microporosity (Ip) in the microcrystalline carbonate mud; SEM image.

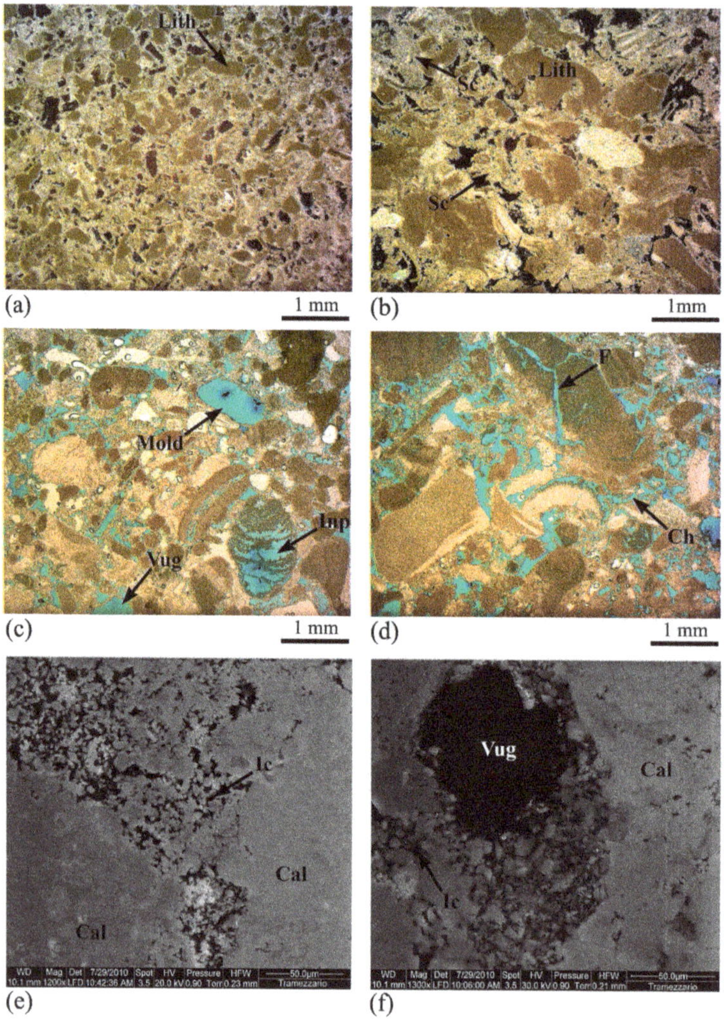

Figure 3. Textural characteristics of Tramezzario: (**a**) Grain-supported texture with bioclasts of Lithothamnium algae (Lith); OM cross polarized light. (**b**) Sparry calcite (Sc) among the bioclasts of Lithothamnium algae (Lith); OM cross polarized light. (**c**) Vug, moldic (Mold) and intraparticle (Inp) secondary porosity; OM plane polarized light. (**d**) Channel (Ch) and fracture (F) secondary porosity; OM plane polarized light. (**e**) Intercrystal (Ic) pores among the calcite (Cal) crystals; SEM image. (**f**) Vug and intercrystal (Ic) porosities and calcite (Cal); SEM image.

Pietra Forte has the characteristics of biohermal–biostroma limestones. This is a rock-preserved biogenic texture, whose original components, mainly Lithothamnium algae (Figure 4a,b), were bound and encrusted together during deposition and stayed substantially in the position of growth. According to the Dunham classification [71], the Pietra Forte is a boundstone, mainly made up of Lithothamnium algae, which embody bioclasts of other organisms such as bivalves, bryozoans, etc., all cemented by sparry calcite, which makes the rock very compact and scarcely porous at the microscopic level. The porosity of Pietra Forte is about 1%–4%, mainly of the secondary type, due to post diagenetic dissolution phenomena. The observed pores are of intraparticle, vug, channel and fracture types (Figure 4c,d). However, pores of the primary type, such as the growth framework due to the growth structures of Lithothamnium algae, have been observed in a few cases. The majority of these pores are now filled with sparry calcite (Figure 4c).

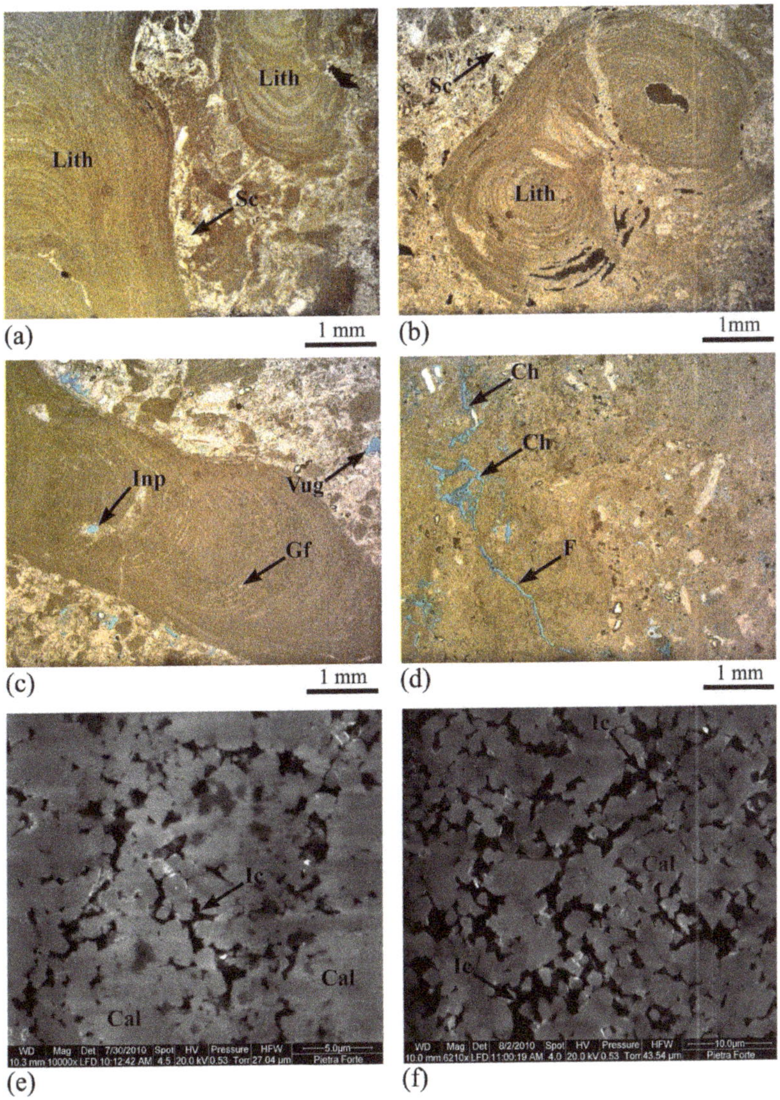

Figure 4. Textural characteristics of Pietra Forte: (**a**) Biogenic texture rich in Lithothamnium algae (Lith) well-cemented by sparry calcite (Sc); OM cross polarized light. (**b**) Another fossil of Lithothamnium algae (Lith) with sparry calcite (Sc); OM cross polarized light. (**c**) Vug and intraparticle (Inp) secondary porosity. The primary porosity due to the growth structure of algae, called the growth framework (Gf), filled by the sparry calcite; OM plane polarized light. (**d**) Channel (Ch) and fracture (F) secondary porosity; OM plane polarized light. (**e**) Intercrystal (Ic) pores among the calcite (Cal) crystals; SEM image. (**f**) Intercrystal (Ic) porosity in a more porous sector of the rock and calcite (Cal); SEM image.

Pietra Forte is exclusively made up of calcite [57], characterized by single subhedral microcrystals (sizes 2–3 μm) or polyhedral aggregates of calcite. With SEM, Pietra Forte is characterized by an intercrystal microporosity [72] with dimensions of pores smaller than 4 μm (Figure 4e,f). This scarce microporosity is due to the presence of carbonate cement, which during lithogenesis filled the spaces between the algal structures and the other bioclasts, making this lithotype very compact. However, when the pores are communicating, they allow the infiltration of water rich in CO_2 that causes dissolution and enlargement of the pores up to the formation of karst cavities.

3.2. MIP Analyses

The results of the MIP study are summarized in Figure 5a–c, where three representative curves showing the typical patterns of the pore-throat size (diameter) distribution of Pietra Cantone, Tramezzario and Pietra Forte are reported. In Figure 5a we can observe that in Pietra Cantone, the sizes of pore-throats are prevalently in the 0.2–7 μm range. In the MIP curve of Tramezzario (Figure 5b), the sizes of pore-throats range mainly from 1 μm to 90 μm. The Pietra Forte representative MIP curve shows pore-throat diameters in the range from about 0.04 μm to 0.4 μm (Figure 5c).

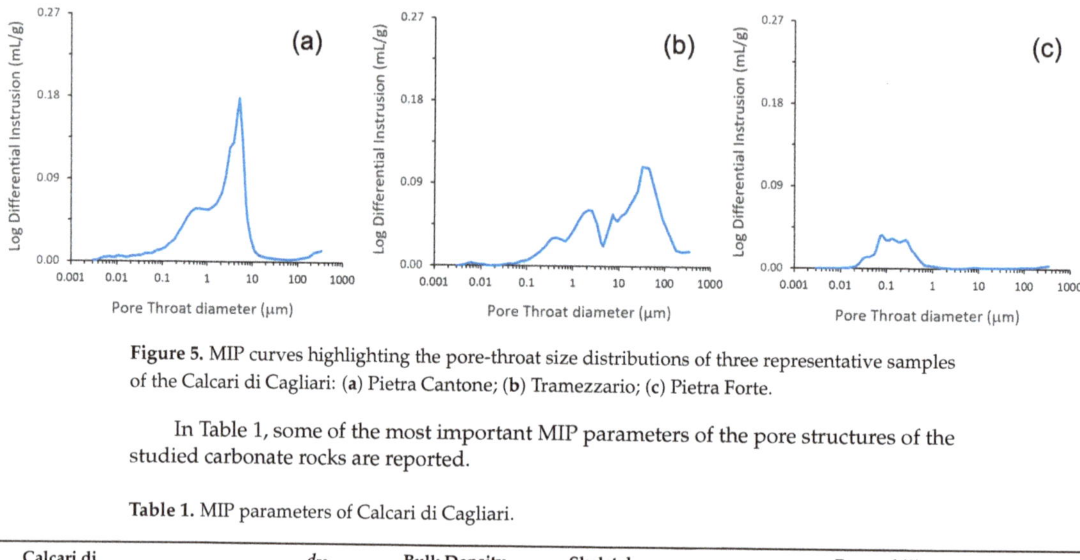

Figure 5. MIP curves highlighting the pore-throat size distributions of three representative samples of the Calcari di Cagliari: (**a**) Pietra Cantone; (**b**) Tramezzario; (**c**) Pietra Forte.

In Table 1, some of the most important MIP parameters of the pore structures of the studied carbonate rocks are reported.

Table 1. MIP parameters of Calcari di Cagliari.

Calcari di Cagliari	Sample	d_{50} (μm)	Bulk Density (g/cm³)	Skeletal Density (g/cm³)	Porosity (%)	Permeability (mD)	Tortuosity
Pietra Cantone	PC1	1.69	2.07	2.77	25.35	8.16	21.23
	PC2	8.85	1.62	2.65	38.71	215.10	5.34
	PC3	1.03	1.89	2.66	28.86	0.37	16.01
	PC4	1.78	1.89	2.72	30.41	9.16	20.20
	PC5	2.45	1.91	2.68	28.72	10.93	23.25
Tramezzario	TR1	12.06	1.58	2.67	40.85	557.92	3.01
	TR2	3.06	2.33	2.69	13.40	26.02	3.73
	TR3	10.67	1.99	2.71	26.61	506.42	4.45
	TR4	17.58	2.09	2.73	23.25	702.28	3.90
	TR5	14.70	1.89	2.75	31.00	1168.12	3.68
Pietra Forte	PF1	0.13	2.53	2.70	6.26	0.11	2.39
	PF2	0.20	2.46	2.71	9.24	0.12	6.55
	PF3	0.15	2.46	2.70	8.74	0.11	6.21
	PF4	16.13	2.51	2.69	6.66	1044.22	2.36
	PF5	0.05	2.65	2.71	2.21	0.16	2.03

d_{50} = median pore-throat diameter.

In the Pietra Cantone samples, the median pore-throat diameters (d_{50}), corresponding to 50% of mercury saturation, range from 1.03 μm to 8.85 μm. They prevail in the range of microporosity (Table 1) according to its textural features (Figure 2a,b). Bulk density values are from 1.62 g/cm^3 to 2.07 g/cm^3 and the values of skeletal density are between 2.65 g/cm^3 and 2.77 g/cm^3. All samples of Pietra Cantone show high values of effective porosity ranging from 25.35% to 38.71%. The values of permeability from 0.37 mD to 10.93 mD in four samples (PC1, PC3, PC4 and PC5) are low (Table 1). On the other hand, except for the PC2 sample (Table 1), the generally high tortuosity values (between 16.01 and 23.25) of these samples are compatible with the intrinsic nature of this lithotype and texture, that is prevalently made up of small-sized particles (micrite matrix) as deduced by OM and SEM. The PC2 has a high permeability value of 215.10 mD (Table 1) that is compatible with its large size of pore-throats (d_{50} of 8.85 μm). In this sample, the MIP results highlight the highest value of effective porosity (at about 40%) related to meso- and macropores (Figure 2c–f) interconnected by larger throats which favor the passage of fluids into the rock. The low tortuosity value of 5.34 is compatible with its high permeability value.

In the Tramezzario samples, the median pore-throat diameters (d_{50}), from 3.06 μm to 17.58 μm (Table 1), are generally higher than the pore-throat sizes measured in Pietra Cantone and Pietra Forte. They prevail in the range of mesoporosity as also highlighted by OM observation (Figure 3c,d). Bulk density values range between 1.58 g/cm^3 and 2.33 g/cm^3 and skeletal density values are from 2.67 g/cm^3 to 2.75 g/cm^3 (Table 1). As for Pietra Cantone, the Tramezzario shows high values of effective porosity (in the 23.25%–40.85% range), except for TR2 sample (13.40%). Permeability values are in the 26.02–1168.12 mD range. The low values of tortuosity, from 3.01 to 4.45 (Table 1), favor the circulation of fluids into the pore network. The generally high permeability and low tortuosity are also related to the interconnected porosity at the meso- and microscale as also observed by OM and SEM (Figure 3c,f).

Pietra Forte limestone shows median pore-throat diameters (d_{50}) from 0.05 μm to 0.13 μm (microporosity), except for the PF4 sample with a median throat diameter of 16.13 μm (Table 1). Bulk and skeletal densities range from 2.46 g/cm^3 to 2.65 g/cm^3 and from 2.69 g/cm^3 to 2.71 g/cm^3, respectively. These samples show values of effective porosity from 2.21% to 9.24% and permeability from 0.11 to 0.16 mD, except for PF4 (1044.42 mD), due to its large pore-throat size. The tortuosity of Pietra Forte is quite low ranging from 2.03 to 6.55 and related to a pore network not articulated depending on the intrinsic textural features of this rock (Figure 4a–f). In general, Pietra Forte is less porous and better cemented than Pietra Cantone and Tramezzario [73] with low values of porosity and permeability. Nevertheless, in some cases (e.g., the PF4 sample), the presence of fractures and dissolution phenomena forming "karst" with caves and fissures can increase its permeability [73].

3.3. CRP

Using the integrated procedure described above, the three high-resolution 3D unstructured filtered and aggregated prismatic-shaped point clouds were computed. They were texturized with the natural colors represented in an RGB scale as detected from the 2D photo images with the aid of the AgisoftMetashape photogrammetric package (Figure 6). These point clouds were subsequently texturized also with the intensity of the light containing the reflectance of the materials making up the samples under study, which can be directly compared with the velocity patterns from the high-frequency ultrasonic methodology.

To account for the accuracy of the HR 3D models computed with the CRP methodology, we first estimated the geometrical anomalies of the point clouds of the Pietra Cantone, Tramezzario and Pietra Forte prismatic-shaped samples with the aid of the RANSAC shape detector utility of CloudCompare. Six planes fitted to the faces of each sample were automatically computed and then meshed. The residuals of the point cloud with respect to the planes using the corresponding meshes as references were computed with the "cloud to mesh" distance (C2M) function of the package. After aggregating the six faces' geometrical anomalies with the merge function of CloudCompare, the statistical Gauss function was subsequently applied in order to estimate the standard deviation of the means of the geometrical anomaly aggregated clouds of the three samples [74]. The results of this last computation are represented in the three histograms of Figure 7.

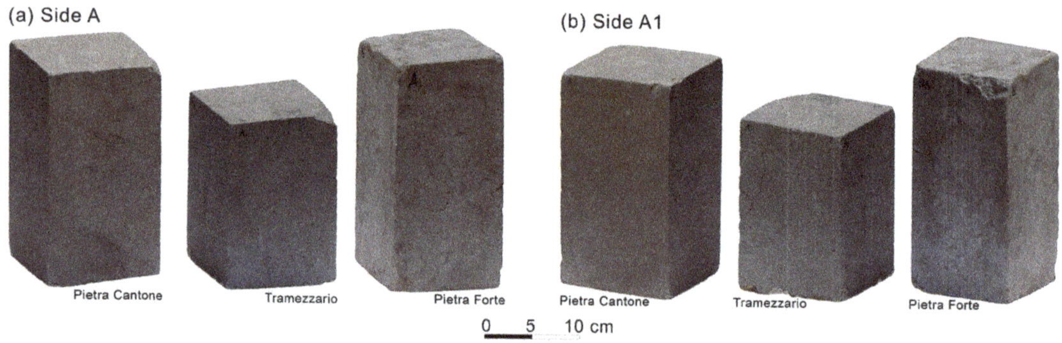

Figure 6. (**a**) Side A view of the HR 3D models of the three prismatic samples under study texturized with natural colors. (**b**) Side A1 view.

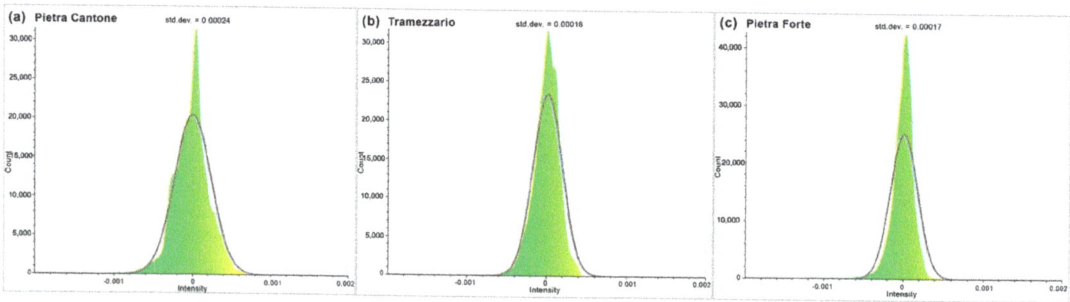

Figure 7. Statistical Gauss function applied to the geometrical anomalies of the three HR 3D models of the prismatic samples under study. The standard deviations of the three models are computed to account for the model accuracy.

As can be appreciated in the figure, the data appear to follow a Gaussian statistical distribution, and the standard deviations of the three 3D models are 0.24, 0.18 and 0.17 mm for the Pietra Cantone, Tramezzario and Pietra Forte sample clouds, respectively, testifying to the great accuracy of the CRP methodology. Finally, considering that for a 3D Gaussian distribution 99% of the confidence region is approximately reached by multiplying the standard deviations by a factor of three, the accuracy of the corresponding 3D models ranges from 0.7 mm for the Pietra Cantone lithotype to 0.5 mm for the Pietra Forte and Tramezzario lithotypes. The greater the accuracy of the model, the smaller the standard deviation. This comparison testifies that according to CRP results, the most compact material of the Calcari di Cagliari formation is Pietra Forte followed by Tramezzario, while Pietra Cantone actually appears to be the least compact.

3.4. Ultrasonic Tomography

The interpretation of the integrated geomatic and geophysical data proposed in this study has been supplemented by petrographical and petrophysical data. The textural and compositional characteristics of the carbonate rocks, as well as the quantity and size of the pores, and the dominant pore types exert a profound influence on the geophysical responses [49,51,75,76]. Therefore, in order to emphasize the benefit that derives from applying the proposed non-destructive multi-technique approach in the characterization of stone materials, it is fundamental to understand the link between petrographical and geophysical properties. Moreover, by understanding the link between petrographic and geophysical parameters, certain ambiguities in geophysical interpretation can be avoided in many practical cases. The composition, texture and type of porosity predispose a stone material to different types and intensities of degradation, which can be recognized and qualified both superficially and in depth by using the geomatic and geophysical methods presented here.

The results of the 3D UT showing high-resolution images of the P-wave velocity distribution on the three significative samples of the Calcari di Cagliari formation are presented together with the 3D CRP high-resolution models and significative thin section images in Figures 8–13. As tomographic results in the above figures, the horizontal tomographic slices extracted from the 3D UT data volume are also reported. The location and orientation of the slices were decided interactively for a detailed mapping of the internal distribution of the elastic characteristics of the investigated samples. As a first result of the ultrasonic evaluation, the velocity ranges were detected for each sample. The velocities range from 2320 m/s to 4390 m/s in Pietra Cantone, from 2750 m/s to 5480 m/s in Tramezzario and from 3500 m/s to 7500 m/s in Pietra Forte. The differences in the longitudinal velocity of the three lithotypes are justified by the complex physical and petrographical properties that characterize the carbonate rocks due to their complex pore system and textural heterogeneities as seen in the results from OM, SEM and MIP.

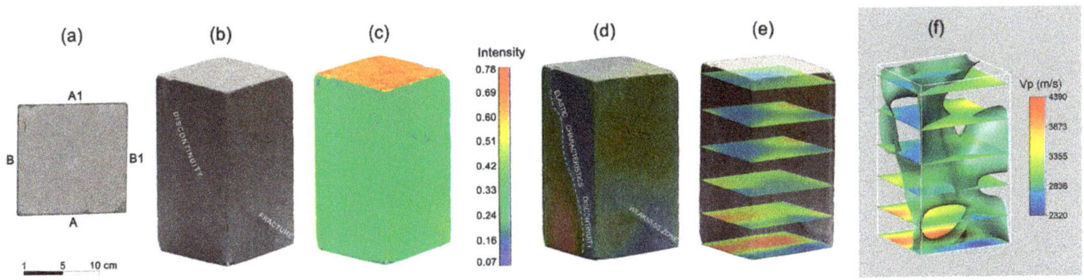

Figure 8. Results of the 3D analyses by the integrated approach for the Pietra Cantone sample. (**a**) Localization of the different faces of the sample. (**b**) 3D HR CRP model texturized with natural colors and represented with an RGB scale. (**c**) 3D HR CRP model texturized with the intensity of ambient light. (**d**) 3D ultrasonic tomography incorporated within the 3D HR CRP model. (**e**) Tomography horizontal slices. (**f**) Isosurface of 2800 m/s with horizontal slices.

As a general remark, it can be observed that the shallow features (e.g., presence of fractures of different origins, discontinuities and chromatic anomalies) of the examined samples as observed in the CRP models (Figures 8b,c, 10b,c and 12b,c) match the longitudinal velocity distribution in the shallow part of the materials (Figures 8d, 10d and 12d), and in many cases they also influence the elastic conditions of their inner parts.

In the Pietra Cantone sample, the mechanical discontinuity and fracture that affect its shallow part, which are highlighted well by the CRP models (Figure 8b,c), determine a significant decrease in the longitudinal velocity (blue zones in Figure 8d,e). Furthermore, the results of the 3D UT indicate that the above mechanical discontinuity evolves into the inner part of the sample as a discontinuity of elastic characteristics along the longitudinal development of the sample. The fracture observable in the CRP model (Figure 8b) does not develop in depth as the low velocity zone is limited to the shallowest parts, as can be observed in Figure 8e. The isosurface of 2800 m/s shown in Figure 8f represents the boundary between high (red) and low (blue) velocity zones. Considering the OM, SEM and MIP results, it can be deduced that discontinuity, fractures and other defects mainly occur in sectors where the rock is most vulnerable on account of a high concentration of secondary porosity of the vug type, which depletes the elastic properties of the carbonate material (low velocity zones) (Figure 9a,b). The vug porosity is probably due to the dissolution processes caused by a permanent presence of water in sectors rich in carbonate mud. As observed by SEM, the mud is characterized by a high percentage of primary micropores of the interparticle type which, when combined, form a porous network that negatively affects the velocity of the ultrasonic pulses, and therefore the elastic characteristics of the rock. However, in the analyzed sample, higher velocity values (approximately 3800–4300 m/s) were also detected. They are related to more compact rock sectors, where the percentage of pores, especially of the secondary type, is low (Figure 9a,c). This condition, due to a better compaction in the diagenetic phase and almost absent dissolution phenomena, enhances the elastic properties of the rock and favors the propagation of the longitudinal ultrasonic pulses.

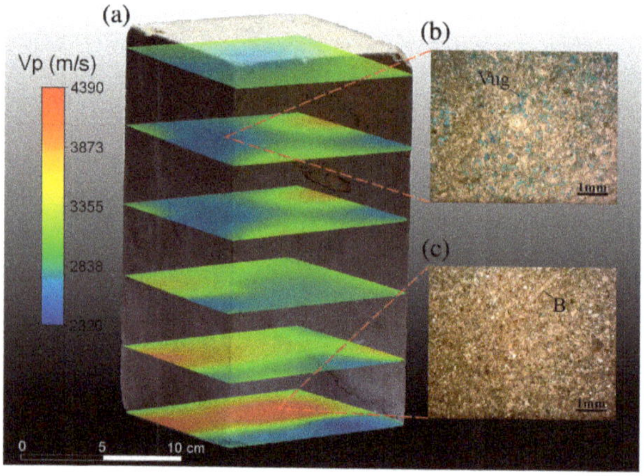

Figure 9. Pietra Cantone sample: integrated analysis of ultrasonic tomography and thin sections. (**a**) Tomography horizontal slices incorporated within the 3D HR CRP model. (**b**) Microscopic features of a porous sector of the sample; OM plane polarized light; thin section treated with blue dye epoxy resin. Vug (vug secondary porosity). (**c**) Microscopic features of a compact sector of the sample; OM cross polarized light. B (bioclast).

In the sample of Tramezzario, the fracture, which is well visible on the CRP models (Figure 10b,c), mainly influences the elastic conditions of the shallow material since it does not develop significantly in depth, as can be deduced from the limited extension of the low velocity zones (blue zones in Figure 10d,e) detected by 3D UT. The isosurface of the 3800 m/s shown in Figure 10f represents the passage between high (red) and low (blue) velocity zones. In the Tramezzario sample, the best cemented sectors, where the deposition of carbonate cement prevailed with respect to the dissolution processes, are characterized by higher velocity values (4700–5480 m/s) typical of a compact carbonate rock with good elastic characteristics (Figure 11a,b).

The lower velocity values (approximately 2750–3500 m/s) detected in a few sectors of the sample are attributable to a greater concentration of secondary pores of channel, vug and fracture types. These secondary pores involve diffraction phenomena causing scattering of the elastic waves, and consequently attenuation of the ultrasonic signal propagation (Figure 11a,c) [77].

In the Pietra Forte sample (Figure 12a–f), which is characterized by higher velocity values compared to previous lithotypes, the weak fracture highlighted by the 3D CRP models does not produce significant variations in the elastic behavior of the rock. This is a consequence of the fact that this fracture is partly cemented and limited to the shallow parts of the material, as can be deduced from the 3D UT results (Figure 12d,e). The isosurface of the 5200 m/s shown in Figure 12f represents the passage between high (red) and low (blue) velocity zones. It can be noted that in some sectors of the analyzed sample, the ultrasonic signals propagate with lower velocity values (3500–4000 m/s). In these sectors, different types of secondary porosity such as channel, fracture and karst microcavities negatively affect the elastic behavior of the rock (Figure 13a,b). The elastic characteristics of this rock are excellent in sectors characterized by high velocity values (6000–7500 m/s). In these sectors, Pietra Forte is almost free of pores and very well cemented by spathic calcite (Figure 13a,c).

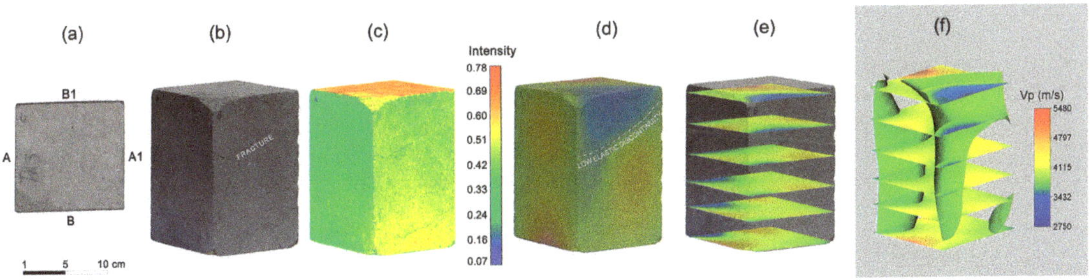

Figure 10. Results of the 3D analyses by the integrated approach for the Tramezzario sample. (**a**) Localization of the different faces of the sample; (**b**) 3D HR CRP model texturized with natural colors and represented in a RGB scale; (**c**) 3D HR CRP model texturized with the intensity of ambient light; (**d**) 3D ultrasonic tomography incorporated within the 3D HR CRP model; (**e**) Tomography horizontal slices; (**f**) Isosurface of 3800 m/s with the horizontal slices.

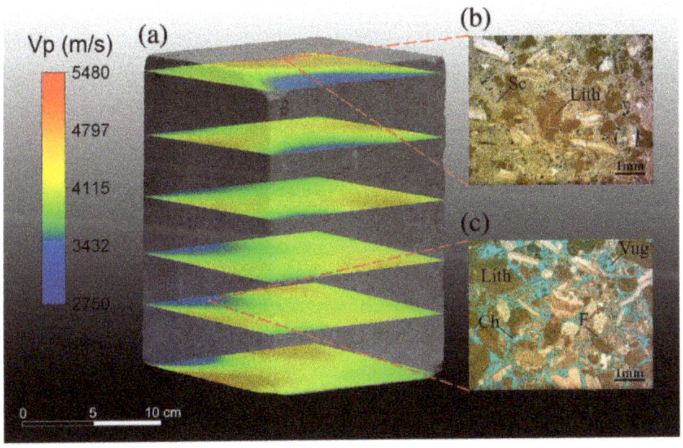

Figure 11. Tramezzario sample: integrated analysis of ultrasonic tomography and thin sections. (**a**) Tomography horizontal slices incorporated within the 3D HR CRP model. (**b**) Microscopic features of a compact sector of the sample; OM cross polarized light. Sc (sparry calcite); Lith (Lithothamnium algae). (**c**) Microscopic features of a porous sector of the sample; OM plane polarized light; thin section treated with blue dye epoxy resin. Vug (vug secondary porosity); Lith (Lithothamnium algae); Ch (channel porosity); F (fracture).

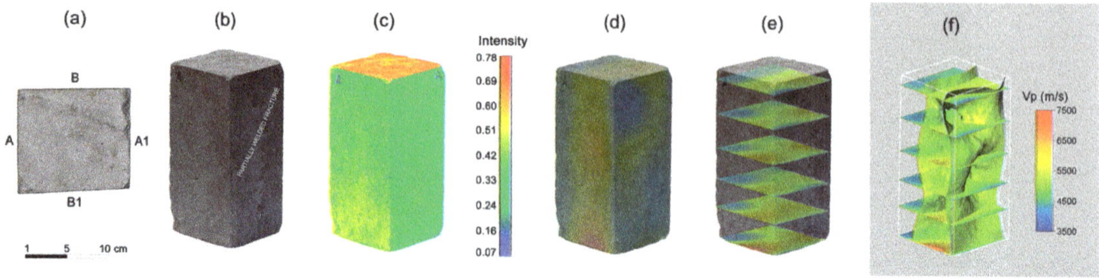

Figure 12. Results of the 3D analyses by the integrated approach for the Pietra Forte sample. (**a**) Localization of the different faces of the sample. (**b**) 3D HR CRP model texturized with natural colors and represented with an RGB scale. (**c**) 3D HR CRP model texturized with the intensity of ambient light. (**d**) 3D ultrasonic tomography incorporated within the 3D HR CRP model. (**e**) Tomography horizontal slices. (**f**) Isosurface of 5200 m/s with horizontal slices.

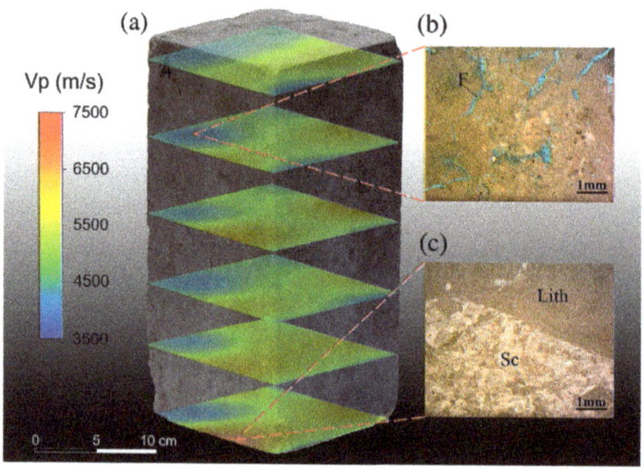

Figure 13. Pietra Forte sample: integrated analysis of ultrasonic tomography and thin sections. (**a**) Tomography horizontal slices incorporated within the 3D HR CRP model. (**b**) Microscopic features of a porous sector of the sample; OM plane polarized light; thin section treated with blue dye epoxy resin. F (fracture secondary porosity). (**c**) Microscopic features of a compact sector of the sample; OM cross polarized light. Lith (Lithothamnium algae); Sc (sparry calcite).

4. Concluding Remarks

The integrated non-invasive approach for the characterization of the carbonate rock samples of the Calcari di Cagliari formation carried out in this study by different techniques can effectively highlight the characteristics of the stone materials and reduce uncertainties that may arise using a single methodology. The integration of different datasets demonstrates that it is capable of solving ambiguities by linking different types of information.

With the aid of close-range photogrammetry (CRP) based on the structure from motion (SfM) methodology, high-resolution 3D models of the studied samples with an accuracy of better than 1mm were computed. For every prismatic sample, a cloud of about 10 million points was computed and texturized, both with the natural colors derived from the 2D digital images and the intensity of the backscattered natural light. The accuracy of the models ranges from about 0.7 to 0.5 mm and was determined starting from the geometrical anomalies computed as residuals with respect to the six planes fitting to the prismatic-shaped sample faces. As a matter of fact, the greater the accuracy the lower the noise due to the presence of defects like voids, cracks and discontinuities of the rock. In fact, the accuracy of the models of the studied samples is greater for Pietra Forte (RMSE = 0.51 mm), while the Tramezzario is characterized by a comparable noise (RMSE = 0.54 mm). The least consistent material appears to be Pietra Cantone that is characterized by an accuracy closer to 1mm (RMSE = 0.72 mm).

Therefore, the CRP methodology is very useful for the analysis of the shallow part of stone materials and provides useful data suitable to different applied research branches. Furthermore, the large amount of acquired metric information was useful for an accurate planning of 3D ultrasonic tomography and for the locations of the ultrasonic sources and receiver stations. The 3D high-resolution CRP models allowed us to reconstruct the ultrasonic survey geometry precisely for an accurate processing of the 3D tomographic data and effective restitution of the tomography results. In fact, the 3D ultrasonic volumes were incorporated within the 3D high-resolution CRP models. This integration became a tool for efficiently understanding the possible links between the conditions of the shallow and inner parts of the sample.

The 3D ultrasonic tomography was effective in efficiently reconstructing a detailed image of the longitudinal velocity distribution inside the samples. This technique, complemented with geomatic and petrographic–petrophysical data, represents a powerful high-resolution method for investigating the inner elastic condition of the stone materials, while identifying heterogeneities, precise locations and sizes of defects or damaged zones at a different scales. From the integrated analysis of the geomatic–geophysical data with the petrographic–petrophysical data, it can be seen that factors such as mineralogical composition, structure and geometry of the pores, type of pores, tortuosity and cementing material need to be taken into consideration while interpreting the geomatic and geophysical responses of the carbonate rocks. Understanding the dependence of the longitudinal velocity on the above rock properties may be important for many practical applications, such as the diagnostic process of the conservation state of monumental structures, rock mass qualification for engineering purposes, quarrying activities, mining and others.

From the integrated analysis carried out in the present study, it appears that the carbonate lithotypes of the Calcari di Cagliari are considerably heterogeneous and characterized by different elasto-mechanical conditions, as demonstrated by the velocity ranges that characterize them. According to their different elastic behaviors, considering the geomatic responses and their petrographical and petrophysical characteristics, it is possible to associate each lithotype with a different susceptibility to degradation.

Pietra Forte is less porous and better cemented than Tramezzario and especially Pietra Cantone. Therefore, Pietra Forte is the highest quality carbonate rock investigated in this study because of its textural and petrophysical properties that enhance its elastic behavior. Among the carbonate lithotypes under study, Pietra Forte appears to be the most mechanically resistant even when used as a construction material, and it ensures better performance than the other carbonate rocks under study. Therefore, this lithotype is in general less susceptible to degradation, except for some situations in which there is a prevalence of a secondary porosity of the vug type as a result of the dissolution process of calcite. In these conditions, the dissolution process can lead to an increase in the size and interconnection of the pores, which can favor its degradation at the microscale giving rise to a karst system at the macroscale.

At the sample scale, the Tramezzario lithotype is unaltered and quite compact as deduced from the geomatic and geophysical responses. Tramezzario is a grain-supported limestone rich in bioclasts cemented by sparry calcite. This characteristic increases its elastic behavior and thus its longitudinal velocity. In the Tramezzario lithotype, generally characterized by well-interconnected macroporosity and microporosity, the process of degradation mainly depends on the easy passage of the fluids (high permeability and low tortuosity) into the well-interconnected pore network. Therefore, in this rock, degradation processes do not occur homogeneously and are generally focused into localized zones depending on the local condition of the pore network.

The most vulnerable lithotype, and therefore most subject to degradation, appears to be the Pietra Cantone lithotype. In this rock, in fact, there were high concentrations of secondary porosity of the vug type, probably caused by the occurrence of dissolution process. Moreover, the diffuse interparticle microporosity concentrated in the fine micrite matrix tends to capture the fluids (e.g., moisture) and retain them in the pore network. Consequently, Pietra Cantone is characterized by high hygroscopicity that favors its subsequent chemical weathering and physical disintegration, as observed when used as a building material. These characteristics worsen the elasto-mechanical behavior of the rock and favor the presence of different types of discontinuities and weakness zones. Therefore, in a framework of a monument protection project, the Pietra Cantone represents the construction material whose conditions need to be constantly monitored over time.

The presented integrated approach aimed at characterize the carbonate rocks of the Calcari di Cagliari formation and evaluate their susceptibility to degradation, can also be useful to determine the effectiveness of restoration actions and to check the compatibility between original and restoration materials.

The results presented in this paper demonstrate that the proposed integrated use of non-invasive complementary methods, such as CRP and 3D ultrasonic tomography complemented, where possible, with petrographic information, would guarantee the reproducibility of the measurements both at the laboratory and field scales, giving valuable information about potential changes in time in the rock condition. The integrated use of CRP and ultrasonic tomography can be considered a preventive and non-invasive conservation tool in line with current trends in the protection of cultural heritage, which require iterative and flexible procedures as opposed to interventional conservation actions, that are generally invasive and more expensive.

Author Contributions: S.F. and G.C. conceived the integrated non-invasive approach; F.C. performed the OM and SEM observations of the materials; G.C. and M.G.B. carried out the CRP survey and processed and interpreted the data; S.F. and F.C. planned and carried out the ultrasonic data acquisition; S.F. processed and interpreted the ultrasonic data and performed the integrated ultrasonic and petrographic–petrophysical data analysis with F.C. and P.L.; P.L. analyzed and interpreted the MIP data. All authors discussed the results and performed the integrated interpretation of the different datasets. All authors contributed to the drafting and revision of the manuscript. All authors have read and agreed to the published version of the manuscript.

Funding: This research was funded by Consorzio Interuniversitario per l'Ingegneria delle Georisorse (CINIGeo), Roma, Italy—research contract n. 376—2023. Resp. Sc.: S. Fais.

Data Availability Statement: Dataset available on request from the authors.

Conflicts of Interest: The authors declare no conflicts of interest.

References

1. Bianco, L. Geochemistry, Mineralogy and Textural Properties of the Lower Globigerina Limestone Used in the Built Heritage. *Minerals* **2021**, *11*, 740. [CrossRef]
2. Dionísio, A.; Martinho, E.; Grangeia, C.; Almeida, F. Examples of the Use of Non-Invasive Techniques for the Evaluation of Stone Decay in Portugal. *Key Eng. Mater.* **2013**, *548*, 239–246. [CrossRef]
3. Lisci, C.; Galhano, C.; Simão, J.; Pires, V.; Sitzia, F.; Mirão, J. Hydrophobic Coatings' Efficiency and Limestones' Resistance to Salt Crystallisation. *Sustainability* **2024**, *16*, 816. [CrossRef]
4. Proietti, N.; Calicchia, P.; Colao, F.; De Simone, S.; Di Tullio, V.; Luvidi, L.; Prestileo, F.; Romani, M.; Tatì, A. Moisture Damage in Ancient Masonry: A Multidisciplinary Approach for In Situ Diagnostics. *Minerals* **2021**, *11*, 406. [CrossRef]
5. Roussel, E.; Vautier, F.; Voldoire, O.; André, M.-F.; Cassar, J.; Fronteau, G.; Phalip, B.; Thomachot-Schneider, C.; Toumazet, J.-P. Quantifying 450 Years of Limestone Weathering Induced by Salt Crystallization on Fortifications in Malta and Gozo. *Geomorphology* **2021**, *378*, 107614. [CrossRef]
6. Sitzia, F.; Lisci, C.; Mirão, J. Building Pathology and Environment: Weathering and Decay of Stone Construction Materials Subjected to a Csa Mediterranean Climate Laboratory Simulation. *Constr. Build. Mater.* **2021**, *300*, 124311. [CrossRef]
7. Aly, N.; Wangler, T.; Török, Á. The Effect of Stylolites on the Deterioration of Limestone: Possible Mechanisms of Damage Evolution. *Environ. Earth Sci.* **2018**, *77*, 565. [CrossRef]
8. Cantrell, D.L.; Hagerty, R.M. Microporosity in Arab Formation Carbonates, Saudi Arabia. *GeoArabia* **1999**, *4*, 129–154. [CrossRef]

9. Fais, S.; Casula, G.; Cuccuru, F.; Ligas, P.; Bianchi, M.G.; Plaisant, A.; Pettinau, A. A Contribution to the Geological Characterization of a Potential Caprock-Reservoir System in the Sulcis Coal Basin (South-Western Sardinia). *Energies* **2019**, *12*, 4524. [CrossRef]
10. Patzek, T.W.; Saad, A.M.; Hassan, A. Multimodal Carbonates: Distribution of Oil Saturation in the Microporous Regions of Arab Formations. *Energies* **2022**, *15*, 1243. [CrossRef]
11. Vázquez, M.A.; Galán, E.; Ortiz, P.; Ortiz, R. Digital Image Analysis and EDX SEM as Combined Techniques to Evaluate Salt Damp on Walls. *Constr. Build. Mater.* **2013**, *45*, 95–105. [CrossRef]
12. Christaras, B.; Auger, F.; Mosse, E. Determination of the Moduli of Elasticity of Rocks. Comparison of the Ultrasonic Velocity and Mechanical Resonance Frequency Methods with Direct Static Methods. *Mater. Struct.* **1994**, *27*, 222–228. [CrossRef]
13. Christaras, B. Non Destructive Methods for Investigation of Some Mechanical Properties of Natural Stones in the Protection of Monuments. *Bull. Int. Assoc. Eng. Geol.* **1996**, *54*, 59–63. [CrossRef]
14. Hrdina, M.; Surový, P. Internal Tree Trunk Decay Detection Using Close-Range Remote Sensing Data and the PointNet Deep Learning Method. *Remote Sens.* **2023**, *15*, 5712. [CrossRef]
15. Puxeddu, M.; Cuccuru, F.; Fais, S.; Casula, G.; Bianchi, M.G. 3D Imaging of CRP and Ultrasonic Tomography to Detect Decay in a Living Adult Holm Oak (*Quercus ilex* L.) in Sardinia (Italy). *Appl. Sci.* **2021**, *11*, 1199. [CrossRef]
16. Bianchi, M.G.; Casula, G.; Cuccuru, F.; Fais, S.; Ligas, P.; Ferrara, C. Three-Dimensional Imaging from Laser Scanner, Photogrammetric and Acoustic Non-Destructive Techniques in the Characterization of Stone Building Materials. *Adv. Geosci.* **2018**, *45*, 57–62. [CrossRef]
17. Cara, S.; Valera, P.; Matzuzzi, C. Morphometric Analysis through 3D Modelling of Bronze Age Stone Moulds from Central Sardinia. *Minerals* **2021**, *11*, 1192. [CrossRef]
18. Casula, G.; Fais, S.; Cuccuru, F.; Bianchi, M.G.; Ligas, P. Diagnostic Process of an Ancient Colonnade Using 3D High-Resolution Models with Non-Invasive Multi Techniques. *Sensors* **2023**, *23*, 3098. [CrossRef] [PubMed]
19. Christaras, B.; Cuccuru, F.; Fais, S.; Papanikolaou, H. Application of Non Destructive Ultrasonic Techniques for the Analysis of the Conservation Status of Building Materials in Monumental Structures. In *Engineering Geology for Society and Territory-Volume 8*; Springer: Basel, Switzerland, 2015; Volume 8, pp. 139–143; ISBN 978-3-319-09408-3.
20. De Fino, M.; Galantucci, R.A.; Fatiguso, F. Condition Assessment of Heritage Buildings via Photogrammetry: A Scoping Review from the Perspective of Decision Makers. *Heritage* **2023**, *6*, 7031–7067. [CrossRef]
21. Dionísio, A.; Martinho, E.; Pozo-António, J.S.; Sequeira Braga, M.A.; Mendes, M. Evaluation of Combined Effects of Real-Fire and Natural Environment in a Building Granite. *Constr. Build. Mater.* **2021**, *277*, 122327. [CrossRef]
22. Fais, S.; Casula, G.; Cuccuru, F.; Ligas, P.; Bianchi, M.G. An Innovative Methodology for the Non-Destructive Diagnosis of Architectural Elements of Ancient Historical Buildings. *Sci. Rep.* **2018**, *8*, 4334. [CrossRef] [PubMed]
23. Brunke, O.; Brockdorf, K.; Drews, S.; Müller, B.; Donath, T.; Herzen, J.; Beckmann, F. Comparison between X-Ray Tube-Based and Synchrotron Radiation-Based μCT. In Proceedings of the Developments in X-ray Tomography VI, San Diego, CA, USA, 10–14 August 2008; Volume 7078, p. 70780U.
24. Kaufhold, A.; Halisch, M.; Zacher, G.; Kaufhold, S. X-Ray Computed Tomography Investigation of Structures in Opalinus Clay from Large-Scale to Small-Scale after Mechanical Testing. *Solid Earth* **2016**, *7*, 1171–1183. [CrossRef]
25. Wolański, K.; Zarudzki, W.; Kiersnowski, H.; Dohnalik, M.; Drabik, K.; Urbaniec, A.; Skupio, R. X-Ray Computed Tomography (CT) Applied for Rock Core Analysis. *Bull. Russ. Acad. Nat. Sci.* **2017**, *5*, 43–50.
26. Omori, T.; Suzuki, S.; Michibayashi, K.; Okamoto, A. Super-Resolution of X-Ray CT Images of Rock Samples by Sparse Representation: Applications to the Complex Texture of Serpentinite. *Sci. Rep.* **2023**, *13*, 6648. [CrossRef] [PubMed]
27. Albertin, F.; Bettuzzi, M.; Brancaccio, R.; Morigi, M.P.; Casali, F. X-ray Computed Tomography In Situ: An Opportunity for Museums and Restoration Laboratories. *Heritage* **2019**, *2*, 2028–2038. [CrossRef]
28. Albertin, F.; Morigi, M.P.; Bettuzzi, M.; Brancaccio, R.; Macchioni, N.; Saccuman, R.; Quarta, G.; Calcagnile, L.; Picchi, D. X-ray Tomography Unveils the Construction Technique of Un-Montu's Egyptian Coffin (Early 26th Dynasty). *J. Imaging* **2022**, *8*, 39. [CrossRef]
29. Ashi, J. 11. Computed Tomography Scan Image Analysis of Sediments1. In *Proceedings of the Ocean Drilling Program, Scientific Results*; Texas A & M University: College Station, TX, USA, 1997; Volume 156, pp. 151–158.
30. Ujiie, K.; Maltman, A.J.; Sánchez-Gómez, M. Origin of Deformation Bands in Argillaceous Sediments at the Toe of the Nankai Accretionary Prism, Southwest Japan. *J. Struct. Geol.* **2004**, *26*, 221–231. [CrossRef]
31. Tonai, S.; Kubo, Y.; Tsang, M.-Y.; Bowden, S.; Ide, K.; Hirose, T.; Kamiya, N.; Yamamoto, Y.; Yang, K.; Yamada, Y.; et al. A New Method for Quality Control of Geological Cores by X-Ray Computed Tomography: Application in IODP Expedition 370. *Front. Earth Sci.* **2019**, *7*, 117. [CrossRef]
32. Martino, D.D.; Bellanova, M.; Cippo, E.P.; Felicetti, R.; Scherillo, A.; Kelleher, J.; Kis, Z.; Gorini, G. A Neutron Diffraction and Imaging Study of Ancient Iron Tie Rods. *J. Instrum.* **2018**, *13*, C05009. [CrossRef]
33. Mongy, T. Application of Neutron Tomography in Culture Heritage Research. *Appl. Radiat. Isot.* **2014**, *85*, 54–59. [CrossRef]
34. Bakirov, B.; Saprykina, I.; Kichanov, S.; Mimokhod, R.; Sudarev, N.; Kozlenko, D. Phase Composition and Its Spatial Distribution in Antique Copper Coins: Neutron Tomography and Diffraction Studies. *J. Imaging* **2021**, *7*, 129. [CrossRef]
35. Snavely, N.; Seitz, S.M.; Szeliski, R. Modeling the World from Internet Photo Collections. *Int. J. Comput. Vis.* **2008**, *80*, 189–210. [CrossRef]

36. Westoby, M.J.; Brasington, J.; Glasser, N.F.; Hambrey, M.J.; Reynolds, J.M. 'Structure-from-Motion' Photogrammetry: A Low-Cost, Effective Tool for Geoscience Applications. *Geomorphology* **2012**, *179*, 300–314. [CrossRef]
37. Adamopoulos, E.; Rinaudo, F. Close-Range Sensing and Data Fusion for Built Heritage Inspection and Monitoring—A Review. *Remote Sens.* **2021**, *13*, 3936. [CrossRef]
38. Mestre-Runge, C.; Lorenzo-Lacruz, J.; Ortega-Mclear, A.; Garcia, C. An Optimized Workflow for Digital Surface Model Series Generation Based on Historical Aerial Images: Testing and Quality Assessment in the Beach-Dune System of Sa Ràpita-Es Trenc (Mallorca, Spain). *Remote Sens.* **2023**, *15*, 2044. [CrossRef]
39. Angelini, A.; Cozzolino, M.; Gabrielli, R.; Gentile, V.; Mauriello, P. Three-Dimensional Modeling and Non-Invasive Diagnosis of a Huge and Complex Heritage Building: The Patriarchal Basilica of Santa Maria Assunta in Aquileia (Udine, Italy). *Remote Sens.* **2023**, *15*, 2386. [CrossRef]
40. Assefa, S.; McCann, C.; Sothcott, J. Velocities of Compressional and Shear Waves in Limestones. *Geophys. Prospect.* **2003**, *51*, 1–13. [CrossRef]
41. Al-Dousari, M.; Garrouch, A.A.; Al-Omair, O. Investigating the Dependence of Shear Wave Velocity on Petrophysical Parameters. *J. Pet. Sci. Eng.* **2016**, *146*, 286–296. [CrossRef]
42. Casula, G.; Fais, S.; Cuccuru, F.; Bianchi, M.G.; Ligas, P.; Sitzia, A. Decay Detection in an Ancient Column with Combined Close-Range Photogrammetry (CRP) and Ultrasonic Tomography. *Minerals* **2021**, *11*, 1114. [CrossRef]
43. Kahraman, S.; Yeken, T. Determination of Physical Properties of Carbonate Rocks from P-Wave Velocity. *Bull. Eng. Geol. Environ.* **2008**, *67*, 277–281. [CrossRef]
44. Kassab, M.A.; Weller, A. Study on P-Wave and S-Wave Velocity in Dry and Wet Sandstones of Tushka Region, Egypt. *Egypt. J. Pet.* **2015**, *24*, 1–11. [CrossRef]
45. Pappalardo, G.; Mineo, S. Microstructural Controls on Physical and Mechanical Properties of Dolomite Rocks. *Rend. Online Della Soc. Geol. Ital.* **2016**, *41*, 321–324. [CrossRef]
46. Anselmetti, F.S.; Eberli, G.P. Sonic Velocity in Carbonate Sediments and Rocks. In *Carbonate Seismology*; Palaz, I., Marfurt, K.J., Eds.; Society of Exploration Geophysicists: Tulsa, OK, USA, 1997; pp. 53–74. ISBN 978-1-56080-038-5.
47. Eberli, G.P.; Baechle, G.T.; Anselmetti, F.S.; Incze, M.L. Factors Controlling Elastic Properties in Carbonate Sediments and Rocks. *Lead. Edge* **2003**, *22*, 654–660. [CrossRef]
48. Fais, S.; Ligas, P.; Cuccuru, F.; Maggio, E.; Plaisant, A.; Pettinau, A.; Casula, G.; Bianchi, M.G. Detailed Petrophysical and Geophysical Characterization of Core Samples from the Potential Caprock-Reservoir System in the Sulcis Coal Basin (Southwestern Sardinia–Italy). *Energy Procedia* **2015**, *76*, 503–511. [CrossRef]
49. Fais, S.; Cuccuru, F.; Casula, G.; Bianchi, M.G.; Ligas, P. Characterization of Rock Samples by A High-Resolution Multi-Technique Non-Invasive Approach. *Minerals* **2019**, *9*, 664. [CrossRef]
50. Soete, J.; Kleipool, L.M.; Claes, H.; Claes, S.; Hamaekers, H.; Kele, S.; Özkul, M.; Foubert, A.; Reijmer, J.J.G.; Swennen, R. Acoustic Properties in Travertines and Their Relation to Porosity and Pore Types. *Mar. Pet. Geol.* **2015**, *59*, 320–335. [CrossRef]
51. Weger, R.J.; Baechle, G.T.; Masaferro, J.L.; Eberli, G.P. Effects of Porestructure on Sonic Velocity in Carbonates. In Proceedings of the 2004 SEG Annual Meeting, Denver, CO, USA, 10–15 October 2004; pp. 1774–1777.
52. Cherchi, A. Appunti Biostratigrafici Sul Miocene Della Sardegna (Italia). *Mém. BRGM* **1974**, *78*, 433–445.
53. Gandolfi, R.; Porcu, A. Contributo Alla Conoscenza Delle Microfacies Mioceniche Delle Colline Di Cagliari (Sardegna). *Riv. Ital. Paleontol. Stratigr.* **1967**, *73*, 313–348.
54. Leone, F.; Pontillo, C.; Spano, C.; Carmignani, L.; Sassi, F.P. Benthic Paleocommunities of the Middle–Upper Miocene Lithostratigraphic Units from the Cagliari Hills (Southern Sardinia, Italy). In *Contribution to the Geology of Italy with Special Regard to the Paleozoic Basement. A Volume Dedicated to Tommaso Cocozza IGCP Project N° 276, Newsletter 5*; Carmignani, L., Sassi, F.P., Eds.; Pacini: Siena, Italy, 1992; Volume 5, pp. 151–158.
55. Cherchi, A.P. *Micropaleontological Researches in Sardinia: Guidebook*; Agip: Milano, Italy, 1985.
56. Barca, S.; Melis, E.; Annino, E.; Cincotti, F.; Ulzega, A.; Orrù, P.; Pintus, C. Note Illustrative Della Carta Geologica d'Italia Alla Scala 1: 50.000, Foglio 557, Cagliari. *Serv. Geol. D'Italia Carta Geol. D'Italia ISPRA.* **2005**, *1*, 240. [CrossRef]
57. Cuccuru, F.; Fais, S.; Ligas, P. Dynamic Elastic Characterization of Carbonate Rocks Used as Building Materials in the Historical City Centre of Cagliari (Italy). *Q. J. Eng. Geol. Hydrogeol.* **2014**, *47*, 259–266. [CrossRef]
58. Étienne, J. Technique d'imprégnation de Roches Par Des Résines Colorées Pour l'étude de La Porosité En Lame Mince. *Rev. L'Institut Fr. Pet. Ann. Combust. Liq.* **1963**, *18*, 611–619.
59. Lønøy, A. Making Sense of Carbonate Pore Systems. *AAPG Bull.* **2006**, *90*, 1381–1405. [CrossRef]
60. Katz, A.J.; Thompson, A.H. Quantitative Prediction of Permeability in Porous Rock. *Phys. Rev. B* **1986**, *34*, 8179–8181. [CrossRef] [PubMed]
61. Katz, A.J.; Thompson, A.H. Prediction of Rock Electrical Conductivity from Mercury Injection Measurements. *J. Geophys. Res.* **1987**, *92*, 599. [CrossRef]
62. Hager, J. Steam Drying of Porous Media. Ph.D. Thesis, Department of Chemical Engineering, Lund University, Lund, Sweden, 1998.
63. Semyonov, D. Agisoft Metashape Manual, LLC. St. Petersburg, Russia. Available online: https://www.agisoft.com/pdf/metashape_2_0_en.pdf (accessed on 27 February 2024).

64. ISRM. The Complete ISRM Suggested Methods for Rock Characterization, Testing and Monitoring: 1974–2006. In *Suggested Methods Prepared by the Commission on Testing Methods*; Ulusay, R., Hodson, J.A., Eds.; International Society for Rock Mechanics; ISRM Turkish National Group: Ankara, Turkey, 2007; p. 628; ISBN 975-93675-4-8.
65. ISRM. Upgraded ISRM Suggested Method for Determining Sound Velocity by Ultrasonic Pulse Transmission Technique. In *Testing and Monitoring: 2007–2014*; Ulusay, R., Ed.; ISRM: Cham, Switzerland; Heidelberg, Germany; New York, NY, USA; Dordrecht, The Netherlands; London, UK, 2014; pp. 95–99.
66. Casula, G.; Cuccuru, F.; Bianchi, M.G.; Fais, S.; Ligas, P. High Resolution 3-D Modelling of Cylinder Shape Bodies Applied to Ancient Columns of a Church. *Adv. Geosci.* **2020**, *54*, 119–127. [CrossRef]
67. Trampert, J.; Leveque, J.-J. Simultaneous Iterative Reconstruction Technique: Physical Interpretation Based on the Generalized Least Squares Solution. *J. Geophys. Res.* **1990**, *95*, 12553–12559. [CrossRef]
68. Mendes, M. A Hybrid Fast Algorithm for First Arrivals Tomography. *Geophys. Prospect.* **2009**, *57*, 803–809. [CrossRef]
69. Fais, S.; Casula, G. Application of Acoustic Techniques in the Evaluation of Heterogeneous Building Materials. *NDT E Int.* **2010**, *43*, 62–69. [CrossRef]
70. Lucia, F.J. Rock-Fabric/Petrophysical Classification of Carbonate Pore Space for Reservoir Characterization. *AAPG Bull.* **1995**, *79*, 1275–1300. [CrossRef]
71. Dunham, R.J. Classification of Carbonate Rocks According to Depositional Textures. *Am. Assoc. Pet. Geol.* **1962**, *1*, 108–121.
72. Choquette, P.W.; Pray, L.C. Geologic Nomenclature and Classification of Porosity in Sedimentary Carbonates. *Am. Assoc. Pet. Geol. Bull.* **1970**, *54*, 207–250.
73. Fais, S.; Cuccuru, F.; Ligas, P.; Casula, G.; Bianchi, M.G. Integrated Ultrasonic, Laser Scanning and Petrographical Characterisation of Carbonate Building Materials on an Architectural Structure of a Historic Building. *Bull. Eng. Geol. Environ.* **2017**, *76*, 71–84. [CrossRef]
74. Elkhrachy, I. 3D Structure from 2D Dimensional Images Using Structure from Motion Algorithms. *Sustainability* **2022**, *14*, 5399. [CrossRef]
75. Garia, S.; Pal, A.K.; Ravi, K.; Nair, A.M. A Comprehensive Analysis on the Relationships between Elastic Wave Velocities and Petrophysical Properties of Sedimentary Rocks Based on Laboratory Measurements. *J. Pet. Explor. Prod. Technol.* **2019**, *9*, 1869–1881. [CrossRef]
76. Verwer, K.; Braaksma, H.; Kenter, J.A. Acoustic Properties of Carbonates: Effects of Rock Texture and Implications for Fluid Substitution. *Geophysics* **2008**, *73*, B51–B65. [CrossRef]
77. Casula, G.; Fais, S.; Ligas, P. An Experimental Application of a 3D Terrestrial Laser Scanner and Acoustic Techniques in Assessing the Quality of the Stones Used in Monumental Structures. *Int. J. Microstruct. Mater. Prop.* **2009**, *4*, 45. [CrossRef]

minerals

MDPI

Article

Conservation Assessment of the Stone Blocks in the Northeast Corner of the Karnak Temples in Luxor, Egypt

Abdelrhman Fahmy [1,2,3,*], Eduardo Molina-Piernas [2] and Salvador Domínguez-Bella [2]

[1] Conservation Department, Faculty of Archaeology, Cairo University, Giza 12613, Egypt
[2] Department of Earth Sciences, Faculty of Sciences, University of Cadiz, Campus Río San Pedro, 11510 Cadiz, Spain; eduardo.molina@uca.es (E.M.-P.); salvador.dominguez@uca.es (S.D.-B.)
[3] Rathgen Research Laboratory, National Museums of Berlin, Schloßstraße 1A, 14059 Berlin, Germany
* Correspondence: abdelrhman.fahmy@uca.es or a.fahmy@smb.spk-berlin.de

Abstract: The Karnak Temples complex, a monumental site dating back to approximately 1970 BC, faces significant preservation challenges due to a confluence of mechanical, environmental, and anthropogenic factors impacting its stone blocks. This study provides a comprehensive evaluation of the deterioration affecting the northeast corner of the complex, revealing that the primary forms of damage include split cracking and fracturing. Seismic activities have induced out-of-plane displacements, fractures, and chipping, while flooding has worsened structural instability through uplift and prolonged water exposure. Soil liquefaction and fluctuating groundwater levels have exacerbated the misalignment and embedding of stone blocks. Thermal stress and wind erosion have caused microstructural decay and surface degradation and contaminated water sources have led to salt weathering and chemical alterations. Multi-temporal satellite imagery has revealed the influence of vegetation, particularly invasive plant species, on physical and biochemical damage to the stone. This study utilized in situ assessments to document damage patterns and employed satellite imagery to assess environmental impacts, providing a multi-proxy approach to understanding the current state of the stone blocks. This analysis highlights the urgent need for a multi-faceted conservation strategy. Recommendations include constructing elevated platforms from durable materials to reduce soil and water contact, implementing non-invasive cleaning and consolidation techniques, and developing effective water management and contamination prevention measures. Restoration should focus on repairing severely affected blocks with historically accurate materials and establishing an open museum setting will enhance public engagement. Long-term preservation will benefit from regular monitoring using 3D scanning and a preventive conservation schedule. Future research should explore non-destructive testing and interdisciplinary collaboration to refine conservation strategies and ensure the sustained protection of this invaluable historical heritage.

Keywords: Karnak Temples; stone blocks; stone decay; stone weathering; quarries; contaminated water; satellite images; conservation

1. Introduction

Stone monuments, which often serve as vital links to our past, face significant threats from a variety of environmental impacts that can accelerate their decay and compromise their archaeological value. Among these, temperature fluctuations, wind erosion, salt attacks, flooding, and earthquakes are particularly detrimental, each contributing to the degradation of these enduring structures in distinct ways. Accordingly, temperature fluctuations can induce thermal stress within stone materials, leading to physical damage, such as cracking and spalling. Repeated cycles of heating and cooling cause expansion and contraction, which can exacerbate existing weaknesses in stone structures [1–4]. For instance, the monuments in Athens and Santorini have experienced deterioration due to the thermal stresses caused by the city's extreme temperature variations [5–7]. Moreover, wind erosion is another critical factor, particularly in arid and semi-arid regions where abrasive windblown particles can wear away the surface of stone monuments over time. This process not only erodes the stone but can also contribute to the loss of surface detail and inscriptions [8–10]. Wind impact is indeed a critical issue in desert regions, like Egypt and Sudan, where natural forces can significantly affect both the landscape and cultural heritage. In these regions, wind-driven erosion poses serious challenges, particularly for historical monuments and archaeological sites [11]. The Sphinx, Pyramids of Giza, and Dahshur Pyramids, for example, have been subjected to significant wind-driven erosion over millennia [12–14].

Flooding introduces a range of destructive processes, including erosion, sediment deposition, and chemical weathering. The interaction between water and stone can exacerbate the degradation through processes, like hydrolysis and salt migration, as seen in the damage to the historic structures of Venice due to periodic flooding [15,16]. Earthquakes pose a unique threat by causing immediate and often severe structural damage. Seismic activity can lead to fractures, collapses, and displacements within stone monuments, compounding existing vulnerabilities [17–20]. Historical records of earthquakes reveal the extent of damage to sites, such as the columns in the Abakainon necropolis and Temple C of Selinunte in Italy and historical minarets in Egypt, highlighting the long-term impacts of seismic events [21,22]. Water infiltration is also a major contributor to stone decay, particularly in areas where rising groundwater or flooding occurs [23]. The infiltration of water into stone structures can lead to a range of problems, including salt migration, hydrolysis, and biological growth [24,25]. The impact of water infiltration on the ancient city of Alexandria has been well-documented, highlighting the complex interactions between water and stone materials [26–28]. In addition, salt attack, often exacerbated by rising dampness, leads to the crystallization of salts within stone materials, causing expansion and internal damage. This phenomenon, known as salt weathering, can significantly degrade the structural integrity of stone monuments [29,30].

In Luxor and the south of Egypt, many pilot studies were conducted for archaeological construction materials' durability, decay, alterations, analysis, and conservation. Rombolà [31] revealed the archaeological risks and loss detection of the Sphinx Avenue in Luxor using GIS, as well as creating future intervention plans for the conservation and management. Fitzner et al. [32] demonstrated the problems of sandstone as the main building material in upper Egypt. The most dominant forms of weathering are loss of materials, soiling, detachment, and deformations. Moreover, Ahmed [33] studied the urbanization change and agricultural activities' impact on Luxor monuments and informed that, after building the high dam, many changes were observed, such as rising water table, soil salinization, and groundwater contamination, which affect the durability of stone monuments in upper Egypt, specifically Luxor. In the same context, Megahed [34] validated the climate change impact on Horus Temple preservation in Edfu. Agricultural and urban expansion led to urban sprawling and rising subsurface levels, which caused damage and decay to the structural and architectural elements, such as stone bleeding, salt crystallization, deformations, and dissolution. Marey [35] studied the pathologies of the wall paintings in the Temple of Thutmosis III at Karnak. He reported that the sodium chloride and other soluble salts caused alterations in the wall paintings. Fahmy et al. [36,37] analyzed the impact of the Nile River on the building material and stability of Osiris Temple down the hill of Bigeh Island. The results demonstrated that the mineralogical content of two various types of sandstone was affected by the water interaction with clay minerals of sandstone, which affected its durability. In addition, they studied and analyzed the construction materials of the Ramses II/Nero Temple in the El-Ashmounein archaeological site, where salt weathering was the greatest damage factor that affected the building materials and their preservation due to the various sources of contaminated water nearby.

The Temples of Karnak, one of the most significant religious complexes in Egypt, still contain hundreds of scattered stone blocks that necessitate thorough study from both archaeological and conservation perspectives. The preservation and revelation of these blocks are crucial for future generations. The blocks may contain inscriptions, carvings, or other forms of ancient Egyptian art and writing that provide invaluable insights into the religious practices, historical events, and daily life of ancient Egypt. In addition, these blocks offer clues about the artistic techniques and stylistic changes over time. This paper focuses specifically on the scattered blocks located in the northeast area of Karnak, in front of the Osiris Chapel, examining conditional and conservation standpoints. To achieve a comprehensive understanding, a multi-faceted approach was employed for condition assessment. This involved in situ observation and documentation of decay and alteration patterns, detailed mapping, and the detection of landscape changes over time using temporal satellite images. Additionally, the hydrological context of the area was analyzed to understand its impact on the stone blocks. This multi-proxy approach not only enhances our understanding of the archaeological significance of the scattered blocks but also informs conservation practices aimed at preserving the Karnak Temples for future generations. By integrating various assessment techniques, this study provides a comprehensive evaluation of the scattered blocks, ensuring that both their historical value and physical integrity are maintained.

2. Architectural and Archaeological Context of the Case Study

The Temples of Karnak, located in Luxor (ancient Thebes, in Egypt), represent one of the largest and most significant religious complexes in the ancient world [38]. Primarily dedicated to the Theban triad of Amun, Mut, and Khonsu, this site offers a rich architectural and archaeological context, reflecting over two millennia of continuous building, renovation, and religious practice. Covering over 100 hectares, the Karnak complex is divided into four main precincts: the Precinct of Amun-Ra, the largest and most significant section centered around the Great Temple of Amun; the Precinct of Mut, dedicated to the goddess Mut and located to the south of the Amun-Ra Precinct; the Precinct of Montu, dedicated to the war god Montu and located to the north of the Amun-Ra Precinct; and the Temple of Amenhotep IV (Akhenaten), located to the east and largely dismantled after his reign [39]. Architecturally, Karnak is renowned for its hypostyle hall, particularly the one built during the reigns of Seti I and Ramesses II, which contains 134 massive columns arranged in 16 rows, with the central columns standing 21 m high. The complex also features ten pylons constructed over different periods, serving as monumental gateways, several obelisks erected notably by Hatshepsut and Thutmose III as solar symbols, and a large artificial lake within the Precinct of Amun-Ra used for ritualistic purposes [40]. Archaeologically, the Temples of Karnak underwent various historical developments. Initial construction likely began in the Middle Kingdom (c. 2055–1650 BCE) with modest shrines and structures. During the New Kingdom (c. 1550–1070 BCE), the majority of the current structures were built, reflecting the zenith of Theban power and the prominence of Amun's cult, with significant contributions from pharaohs, such as Thutmose I, Hatshepsut, and Ramesses II. From the Late Period to the Ptolemaic Period (c. 664–30 BCE), the complex saw continued expansions and restorations, including contributions from Nubian pharaohs and Ptolemaic rulers. These phases of construction and renovation highlight Karnak's enduring religious and cultural significance throughout ancient Egyptian history [41].

3. Geographical and Geological Context

The Temples of Karnak are situated in the upper area of Egypt/Luxor city with the coordinates N 25°43′7″, E 32°39′31″ (Figure 1a,b). The total area of Karnak is around 279.48 km^2 and the total area of the case study in the northeast corner of the Karnak Temples is around 3.6 km^2 (Figure 1c). The case study is located around 200 m from the surrounding village of Karnak and around 1 km from the Nile River (Figure 1c). This complex is located on the eastern bank of the Nile River and it is considered an enormous open-air museum and the biggest ancient Egyptian religious community globally.

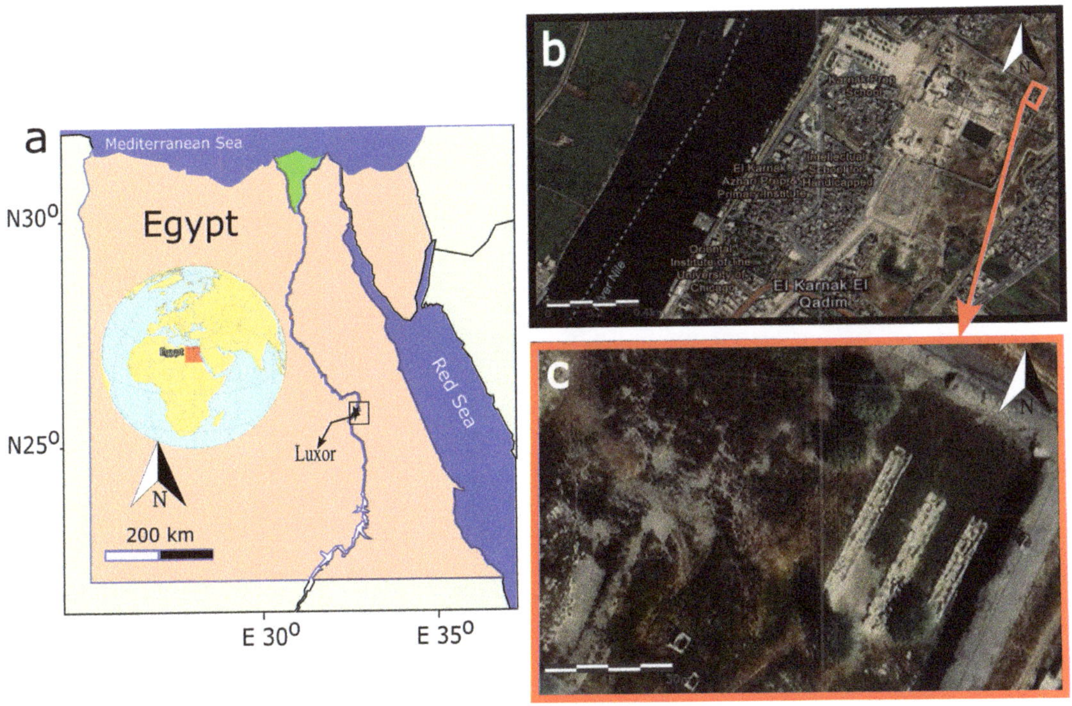

Figure 1. (**a**)Map of Egypt showing the general location. (**b**) Hybrid imagery of the Karnak Temple and its surroundings, including rural houses, agricultural lands, and the Nile River. The case study area is marked by a red square in the northeast corner of the Karnak Temples (**c**).

Geological features of middle and upper Egypt date back to the Eocene age from sedimentary formations (sandstones/siltstone and chalky limestones) with low relief topography (Figure 2) [42]. Luxor geology comprises various units that go back to the Upper Cretaceous, Paleocene–Eocene, Pliocene, and Pleistocene-Holocene [43]. The Pliocene unit is composed of Paleonile river sediments (intraformational conglomerate, red-brown clays with thin fine-grained sand and silt laminae). The Pleistocene unit consists of the Protonile sediments from the Early Pleistocene (pebbly and bouldery gravels), the Prenile sediments of the Middle Pleistocene (fluvial sands with minor gravel and clay beds and loosely consolidated gravel of polygenetic origin), and the Neonile sediments of the Late Pleistocene age (Figure 2). The Holocene unit is formed by the Neonile sediments (silt, micaceous sand, and clays) [43]. The Upper Cretaceous–Early Eocene rock units are unconformably overlain by the Pliocene–Holocene units stratigraphically [44]. The main geological setting of the Karnak Temples is Nile Holocene silty clay along with younger alluvial plains (Figure 2). In addition, the elevation of the Theban plateau is around 300 m above sea level with a sharp scarp encountering the Nile valley on its eastern side (Figure 3) [42]. Around 3.17 km from Karnak in new Luxor city, the soil was characterized by [44]. The mineralogical analysis confirmed that the soil contains smectite minerals as major clay minerals and other clay minerals are present as well, such as kaolinite, chlorite, and illite. In this sense, the presence of such clay minerals can cause a lot of damage to the monuments, bearing soil, foundations, and structural elements because of swelling and shrinkage behaviors.

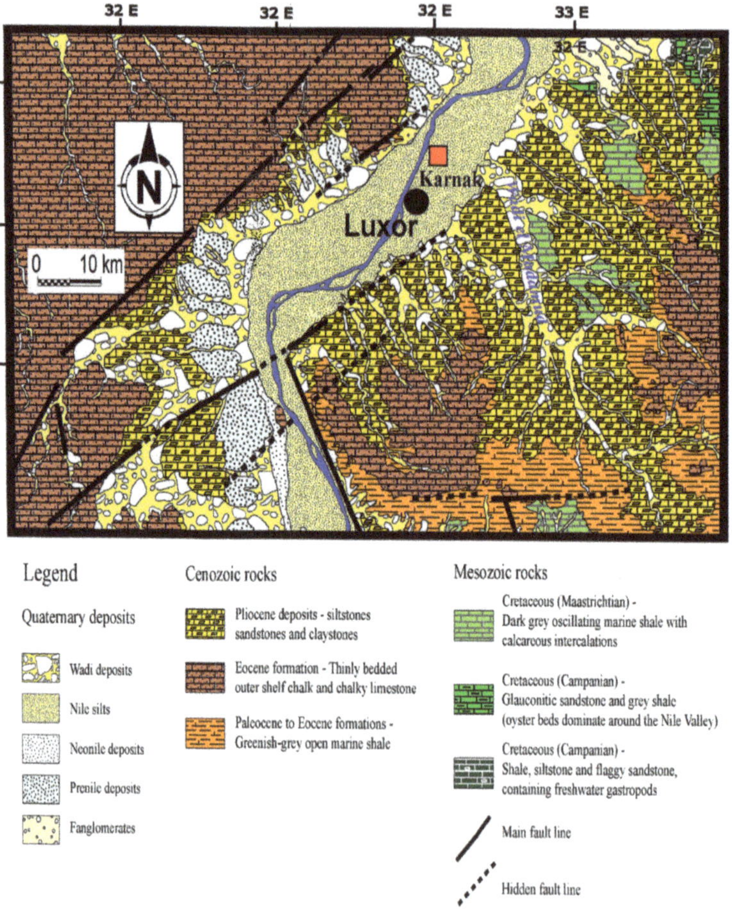

Figure 2. Geological map of Luxor and the Theban Plateau, highlighting the Temples of Karnak and the case study area marked by a red square. Modified after [42].

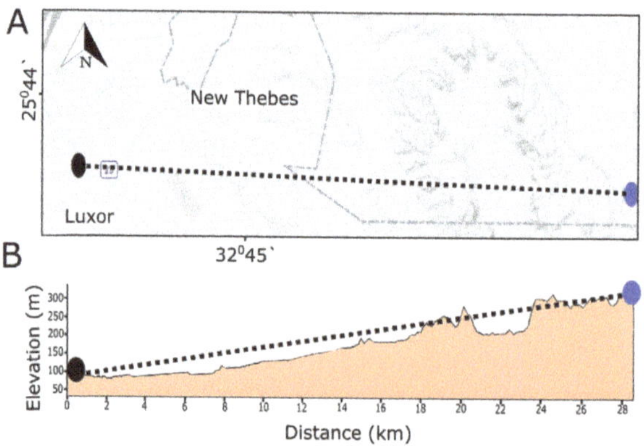

Figure 3. (**A**) Base map and (**B**) elevation profile showing the elevation levels of carbonate formations in the Theban area. The black circle marks a point within the Karnak Temples and the blue circle marks a point approximately 28 km to the east.

4. Quarries of Stone Blocks and Mineralogical Characterization

Generally, ancient Egyptians experienced mining and quarrying and they utilized various kinds of stones for structural, architectural, and decorative purposes. Limestone, sandstone, granitoid, basalt, and alabaster/travertine were prepared and cut from several quarries and locations in Egypt [45–47]. These natural stones were used as building and construction materials in the pyramids, temples, and tombs [46]. The main construction material of Karnak monuments is sandstone with high porosity, especially for the stone blocks in the northeast corner. However, red granites, quartzites, limestones, and travertine/Egyptian Alabaster were observed as building materials as well, in a low quantity. Predominantly used sandstone was extracted mainly from Gebel el-Silsila formations (one group of major Nubian Sandstone), which extend from Edfu to Kom Ombo. All sandstone quarries, except the one from Gebel el-Ahmar in Cairo, date back to 65.5–145.5 million years (Cretaceous period) [45]. The ancient name of Gebel el-Silsila is Kheny and this name disappeared in the third intermediate period but in the Roman period was called 'Sil-sil'/'Silsilis' and many outcrops on the east and west banks of the Nile were exploited in the late dynastic to early Roman Period [48–50]. For instance, in the Ramesside era, quarrying activities were very active and intense, especially in the 19th dynasty. Around three outcrops have been confirmed as quarrying areas on the east bank of the Nile River. Moreover, on the west bank (22nd dynasty), quarrying activities were recorded. On the other hand, the east bank (26th dynasty) witnessed quarrying works as well. In the Ptolemaic period, new sandstone blocks were prepared from the west and east banks of sandstone outcrops for temple reconstruction due to interventive restoration works. Furthermore, in the Roman period, various quarries in Gebel el-Silsila were intensively recoursed. Around 104 quarries were used in the Roman period and 36 outcrops are assured to have been utilized by Romans on the west and east banks of the Nile River (for example, in Edfu, Dendara, Esna, Medamud, the Temple of Isis, and Min at Koptos) [48]. Two indicators can differentiate between dynastic and Roman quarrying for structural and architectural block preparations. Dynastic blocks are much bigger than Roman ones. Nevertheless, Roman block dimensions are 55–60 cm (in height and depth) and 120–290 cm (in width). In dynastic eras, bronze and copper tools were used for block preparations but Romans were using tools that were made of iron and the tool marks are characterized to be systemic and powerful. Additionally, waste left in the field was much more common in the Roman period than in dynastic eras [48]. Harrel and Storemyr [45] mapped the ancient quarries in the west bank from Naq el-Fugani to Gebel Gulab and they classified the silicified sandstone according to their utilities. For example, there were silicified sandstone for prehistoric tools, grinding stones, and ornamental and building stones. Four samples of sandstone were collected to be classified and studied petrographically by [32]. The first sample (east bank) is a light brown spotted fine-grained sandstone with mainly quartz and secondary constituents, such as rock fragments, clay minerals (kaolinite and clinochlore), opaque matter, feldspar (microcline and albite), muscovite, and heavy minerals with low mechanical strength. The second sample (east bank) is a white spotted fine-grained sandstone with mainly quartz and secondary constituents, such as rock fragments, clay minerals (kaolinite and clinochlore), opaque matter, albite, muscovite, heavy minerals, and small amounts of gypsum and calcite with low to medium mechanical strength. The third sample (west bank) is a yellow-brown very-fine-to-five-grained sandstone with mainly quartz and secondary constituents, such as rock fragments, clay minerals (kaolinite and clinochlore), opaque matter, feldspar (microcline and albite), mica (muscovite, biotite), and heavy minerals with low mechanical strength. The fourth sample (east bank) is a white fine-grained mixture with mainly quartz and secondary constituents, such as rock fragments, clay minerals (kaolinite and clinochlore), opaque matter, and heavy minerals with low mechanical strength [32].

5. Hydrological and Climatological Context

The Temples of Karnak and its stone blocks are surrounded by various water sources, sometimes with contaminated water, that infiltrate the stone and structures through infiltrative phenomena to reach building materials and the sources of water can be the Nile River, canals, drains, aquifers, agricultural activities, and urban waste (Figure 4). A group of researchers collected water samples from the Nile River, nearby canals, agricultural areas, and the sacred lake of the Memnon Temple in the west bank/Luxor to check and analyze the salinity of the water chemically [51]. Results showed the presence of chemical contaminants, such as Sodium (Na^+) (195 ppm in Memnon Temple Lake and 36 ppm in the Nile River), Potassium (K^+) (3.5 ppm in Memnon Temple Lake), Sulfate ($SO_4{}^{-2}$) (96 ppm in Memnon Temple Lake), and Chloride (Cl^{-1}) (21 ppm to 190 ppm in Memnon Temple Lake). The possible sources of these chemical constituents are mainly from agricultural irrigation activity and soil dissolution. Subsurface water and high levels of groundwater in the Temples of Karnak were observed between the 1990s and 2006. The water level fluctuations led to the absorption of saline solution triggering the material decay, mechanical decohesion, and subsequent structural instability. In this regard, a groundwater-lowering project and monitoring were conducted from 2007 to 2012, funded by USAID [52]. However, in the summer of 2023, the water rising was observed, also affecting the stone blocks in the Temples of Karnak. Furthermore, the Temples of Karnak are located on two main groundwater aquifers; the shallow Quaternary aquifer (graded sand and gravel with clay with 5 to 95 m thickness and an east to west flow direction) and the Plio-Pleistocene aquifer (sand, clay, and gravel; high salinity) [53]. In this context, groundwater flows from adjacent agricultural areas passing through the Karnak Temples towards the Nile River and the salinity of the groundwater escalates from east to west towards the site and Nile River (Figure 4). Salts trapped within the silty clay led to the increase of salinity under the temples. Additionally, various water samples were collected and analyzed. The results demonstrated that the salinity in the area is very different depending on the site: in the Nile water, it is 173 mg/L; in the canal water, it is 280 m/L; in the drain water, it is 426 mg/L; the water of the sacred lake in Karnak is 780 mg/L; and the groundwater salinity beneath the Karnak Temples ranges from 524 mg/L to 1363 mg/L [53]. The summarized hydrological data context is tabulated in Table 1.

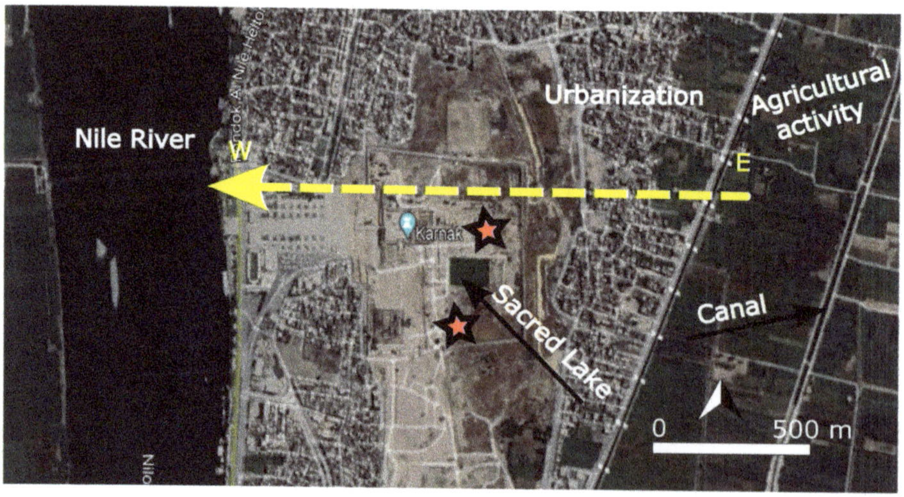

Figure 4. Various contaminated water sources in and around the Temples of Karnak. The yellow arrows indicate the flow direction of water from east to west, originating from canals, drains, cultivated lands, and houses. Salt-contaminated aquifers beneath the Temples of Karnak are marked with stars [53].

Table 1. Summary of hydrological and climatological data for the Karnak area.

Hydrological Context	**Details**
Main Water Sources	Nile River, canals, agricultural areas, drains, aquifers, urban waste.
Contamination Sources	Agriculture (irrigation), soil and structure dissolution, groundwater salinity.
Water Contaminants	Sodium (Na^+), Potassium (K^+), Sulfate ($SO_4{}^{-2}$), Chloride (Cl^{-1}).
Salinity Levels	- Nile River: 173 mg/L - Canal: 280 mg/L - Drains: 426 mg/L - Sacred Lake: 780 mg/L - Groundwater beneath Karnak: 524–1363 mg/L.
Groundwater Aquifers	- Shallow Quaternary aquifer (5–95 m thickness) - Plio-Pleistocene aquifer (high salinity).
Water Movement and Impact	Groundwater flows from agricultural areas eastward toward the Nile River, increasing salinity levels beneath Karnak.
Rising Groundwater Issue	Periodic water level rise observed since the 1990s. Groundwater-Lowering Project (2007–2012) was conducted but water levels increased again in 2023.
Climate Context	**Details**
Summer Temperatures	Ranges from 39 to 44 °C (June–September), with very low humidity and almost no rainfall (1–2 mm).
Winter Temperatures	Ranges from 23 to 35 °C (November–April) with cooler nights having temperature drops of 5 °C. Slight rainfall of up to 4.2 mm in May and 0.0 mm in August.
Climate Projections (2100)	Temperature increases of 3–3.5 °C are expected. Rainfall decreases of −5% to 10% are expected.
Wind Conditions	Winds blow from the east. Daily wind speed ranges from 1 m/s to 5 m/s. March 2024 wind speeds: max 7.19 m/s, min 1.02 m/s.

Luxor is dry with low humidity, extremely hot in the summer, and moderately cold in the winter. In the summer (June–September), the temperatures range between 39 and 44 °C. The precipitation is very rare to absent, with around 1–2 mm of rainfall (May–October). On the other hand, in the winter (November–April), the temperatures are between 23 and 35 °C and nighttime is cold, with the temperature being 5 °C lower. Rainfall reaches 4.2 mm in May and 0.0 mm in August [54]. According to future trends and climate change scenarios, the temperature in Luxor will be increased by around 3 to 3.5 by 2100. Additionally, the rainfall will be decreased by −5% to 10% [54]. The wind direction is East and wind speed varies daily from 1 m/s to up to 5 m/s. On 9 March 2024, the maximum speed of wind was 3.08 m/s and the minimum speed was 1.02 m/s. On 10 March 2024, the maximum speed was 7.19 m/s while the minimum speed was 1.02 m/s (https://meteologix.com/site/selection; accessed on 1 July 2024). The summarized climatological data context is tabulated in Table 1.

6. Archaeoseismic and Flooding Hazard

Earthquakes and flooding events were considered geohazards that affected the structural stability of the Temples of Karnak. Luxor and its monuments are located around an effective seismic region in south Egypt, especially in Aswan, which experiences many earthquakes, such as being the epicenter in the Aswan area of the 1981 earthquake with 5.6 M, as well as Abu Dabbab (Red Sea) in 1955 with 5.6 M and in 1984 with 5.1 M [55]. Ancient earthquakes of 27 B.C, 600 B.C, and 1200 B.C caused a significant amount of damage to Thebes [56]. In this context, El-Sayed et al. [57] added that Aswan and Kalabsha are considered the most active zones and on March 1969 a very strong earthquake happened with 6.9 M at the entrance of the Gulf of Suez that was felt in and outside Egypt and caused strong damage. Moreover, Karakhanyan et al. [56] carried out a structural analysis for

Colossi of Memnon looking at the evidence of seismic effect and its damage. Structural and ongoing cracking and splitting of the colossus and the pedestals were caused due to seismic events, soil deformation/liquefaction, and subsidence. On the other hand, floods greatly affected the stability of most Egyptian monuments, including the Luxor and Karnak Temples, causing structural deformations and destruction. Before 1960, floods were occurring from June to September [58]. In 1898, B.W Kilburn photographed the first pylon and the forecourt area that was extremely flooded by inundation (Figure 5) [59]. In addition, the Temple of Amenhotep III was flooded with approximately 3 m of the flood basin in the Ramesside period, causing the destruction of the temple. However, earthquakes participated greatly in the destruction of the temple as well [60]. In the same context, Mclane et al. [61] reported that around 24 historical flash floods damaged and caused decay for the Valley of Kings tombs in the west bank/Luxor. In Karnak, the flooding continued and its severity levels changed from 1650 B.C. to the construction of the first Aswan Dam. When Karnak was flooded in a specific area, it returned to its state before construction and that means full destruction. The revealed stela in the third pylon in the Temple of Amon-Re at Karnak led to reports about a great flooding that occurred and inundated the Weskhet-hall of the temple [62].

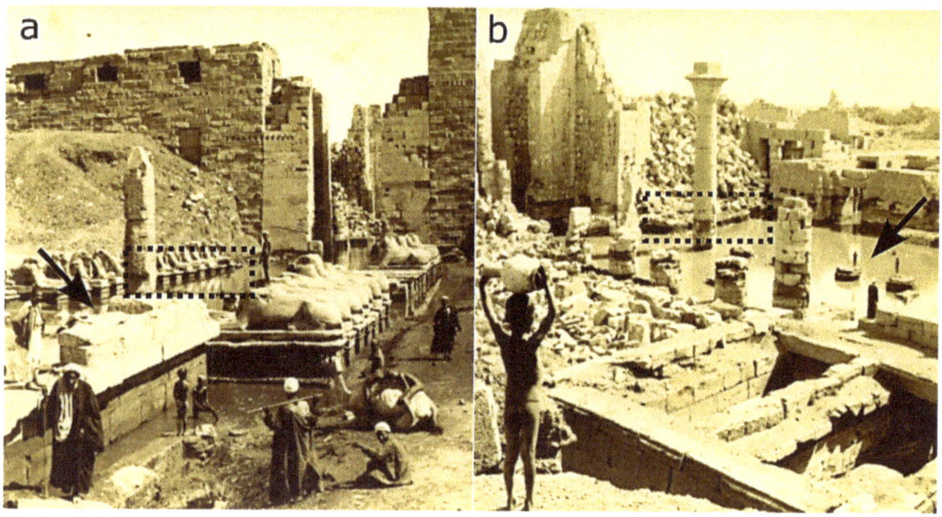

Figure 5. Two historical photographs from 1898 by B.W. Kilburn showing the first pylon of the Karnak Temples. The stone building materials are visibly affected by flash flooding, as indicated by black arrows and highlighted within dotted rectangles (**a**,**b**). From [59].

7. Methodology

To thoroughly study and assess the preservation state of the stone blocks in the Karnak Temples, particularly in the northeast corner, an extensive and multi-faceted research methodology was employed. Field trips were conducted to facilitate comprehensive recording, photographing, and measurement of the stone blocks' condition. These field activities were complemented by detailed desk work, which included mapping, engineering, drawing, and site elevation documentation. In addition, satellite imagery and observation played an important role in this study to detect the vegetation change. It was utilized to specify the distribution of the neglected stone blocks across the entirety of the Temples of Karnak, supplementing the insights gained from on-site visits. Furthermore, insights on the quality of ancient quarry stones as sources of Karnak construction materials were discussed.

For land monitoring and biomass change detection in Luxor and the study area within Karnak, the Copernicus collection of 300 m NDVI V2 was employed. This involved synthesizing images over 10-day intervals based on maximum value composition, allowing for precise tracking of vegetation and landscape changes over time. The NDVI V2 time series was derived from the latest reprocessed data from Sentinel-3/OLCAI orbital segments, which provide daily Top of Canopy (TOC) surface reflectance. The 300 m NDVI V2 processing chain included the use of Aerosol Optical Thickness (AOT) data sourced from the Near Real-Time (NRT) database provided by the Copernicus Atmosphere Monitoring Service (CAMS). This comprehensive approach ensured accurate and up-to-date information on vegetation and environmental conditions affecting the stone blocks. This multi-dimensional assessment methodology, combining fieldwork with advanced satellite data analysis, allowed for a thorough evaluation of the preservation state of the stone blocks. By integrating direct observations with remote sensing technologies, this study not only identified the current conditions and distribution of the blocks but also provided valuable insights into the environmental factors influencing their preservation. This integrated approach supports the development of effective conservation strategies to safeguard the stone blocks of the Temples of Karnak for future generations.

8. Results and Discussion

8.1. In Situ Assessment and Stone Decay

Field visits and present work were conducted to survey and monitor the locations and condition of all stone blocks in the Temples of Karnak and specify their places and problems. The study area has more than 200 mastabas that carry thousands of dismantled stone blocks. These stone blocks have various dimensions and sizes that range between 100 kg and more than 5 tons. Additionally, these stone blocks were prepared and cut from different materials, such as sandstone (major), limestone, red granite, quartzite, etc. These blocks were parts of the constructed temples, chapels, pylons, and other buildings. Some of these blocks still have aesthetical and epigraphical importance and pigments. From visual observation, conservation, archaeological and management plans, and studies are needed to preserve and reallocate them in harmony with the monuments of Karnak. The positions and places of these forgotten stone blocks were mapped on the master plan of the site (Figure 6). They are distributed and focused in the north, west, and center of the Temples of Karnak. In addition, these blocks were classified into six areas or zones (Figure 6). Google Earth satellite images were prepared to show the actual condition and its situation inside the Karnak complex (Figure 7). In this regard, it is observed that the stone blocks in Areas 1, 2, 3, and 5 are positioned randomly. Figure 6b shows the current landscape of the study area. It consists of three big mastabas and they are marked as m1, m2, and m3, respectively. The m1 contains 59 stone blocks and m2 carries 41 stone blocks while m3 has 43 pieces. Some mastabas were constructed from red brick, filled with sand and rock fragments, and then covered by a cement layer. Additionally, some mastabas were constructed from concrete. The bearing capacity of these mastabas to the stone blocks is insecure and incapable of its functionality. Many fractures, cracks, and deformations were observed. Subsurface water and vegetation damaged, physically and mechanically, these mastabas intended to carry tons of valuable stone blocks. An enormous number of stone blocks were found dismantled and scattered mechanically. In addition, physical and chemical weathering and alterations were observed due to various geoenvironmental and inner defects of natural stone.

Mechanically, seismic and flooding events affected the durability and persistence of the Temples of Karnak throughout the ages. Thousands of scattered and deformed structural and architectural elements were observed due to the accumulative impact of earthquakes and annual flash flooding (June–September) or anthropogenic actions until 1960. Many columns of the temples were dismantled and affected, especially at the bottom of columns that can be damaged because of the high straining action (lateral loads of earthquakes and Nile flooding). The response of using stone blocks as structural materials to earthquakes was weak, especially with high degrees of ground motion. The stone unities can be separated, deformed, and damaged into scattered pieces with partial and total failure (Figure 8). Earthquakes can cause out-of-plane displacements that lead to severe damage to temple walls due to macroseismic or accumulative microseismical events. Accordingly, roofs, columns, and corners will be affected by various deformations and dismantling (Figure 9(1a,1b)). The damage to building blocks is directly related to primary faulting when the ground transmits the seismic waves from the focus point to the building through the soil and basements [63]. In this sense, Yön et al. [64] demonstrated that out–in plane movements occur during earthquakes. Shear forces and loads for columns, beams, and joints can be increased with seismic events. In addition, columns without a transverse reinforcement in the plastic hinge region manifest low efficacy and capacity towards lateral dynamic loads. Thus, disfiguring and collapse could be observed. In the same context, Kázmér [18] illustrated that chipping, as a damage aspect, can be caused by seismic or/and weathering actions. To differentiate, chippings from seismic events have sharp edges while chippings from weathering are somehow rounded and peeled. Furthermore, unexpected earthquakes occur via connected or/and unconnected fractures through a single or more than two blocks due to the impact of the vertical part of the seismic wave that shakes the blocks up and down rapidly. Additionally, various oriented cracks can be present in a single block or multiple blocks depending on the stone setting. The Temples of Karnak and buildings are settled on silty muddy soil and are considered a weak point and base that accelerate the building collapse and damage.

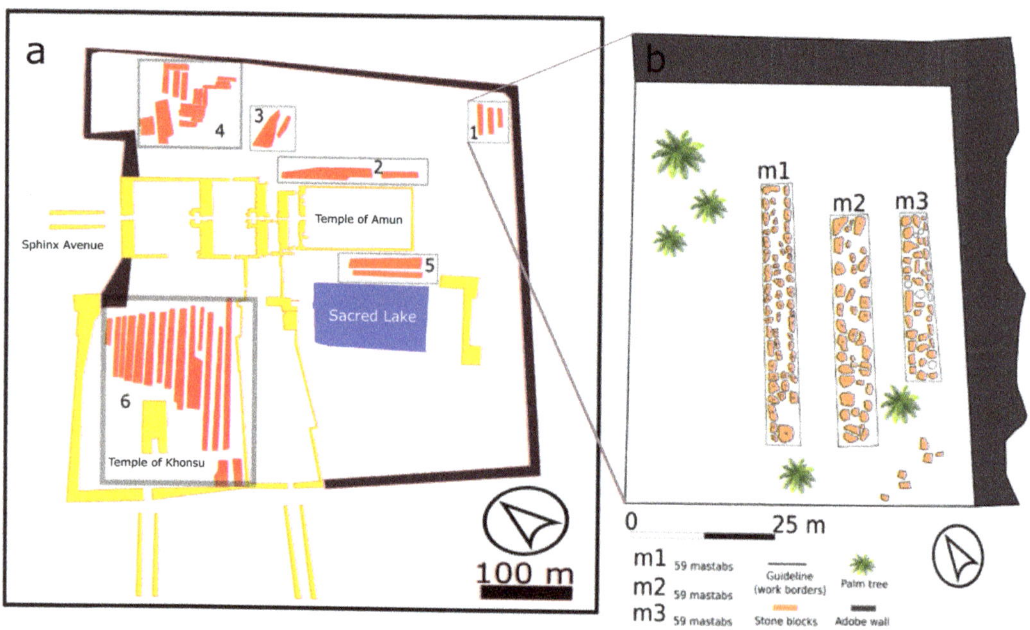

Figure 6. Master plan for the Temples of Karnak. (**a**) Scattered stone blocks are (in red), divided into six areas. (**b**) Case study area in the north east corner of the site, with current archaeological landscape.

Additionally, groundwater and water source levels pose significant threats in Karnak. In terms of this concern, soil types, especially clay soils, are very sensitive and mislay their mechanical strength due to the dynamic loads and forces. In the presence of water, soil liquefaction can happen during the earthquake occurrence. This soil liquefaction can cause sliding and collapse of the buildings [65]. The Temples of Karnak and, in the case study, the stone blocks demonstrated damage evidence from earthquakes' impact. For instance, the decay forms are block fracturing, sharp chippings (dipping broken corners), cracking, and separation (Figure 9(1a,1b,1f,2a,2b)). Fracturing and chipping are the predominant damage aspects in the stone blocks. Stone blocks slipped and embedded into the clay soil are observed as well (Figure 9(1a,1c)).

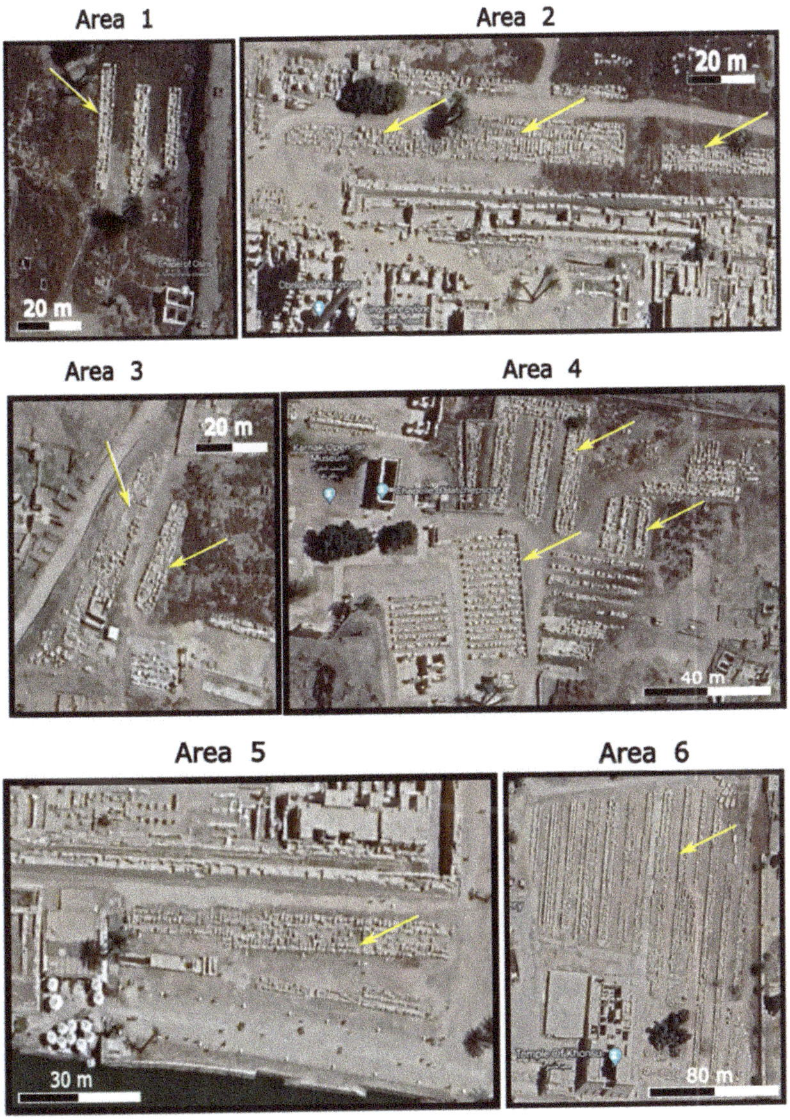

Figure 7. Satellite images depicting six different areas within the Temples of Karnak. The current positions of stone blocks are highlighted by yellow arrows in each image, providing a clear view of their distribution and placement within the temple complex. These images offer insight into the spatial arrangement and condition of the stone blocks, as observed in the present day (Areas 1–6).

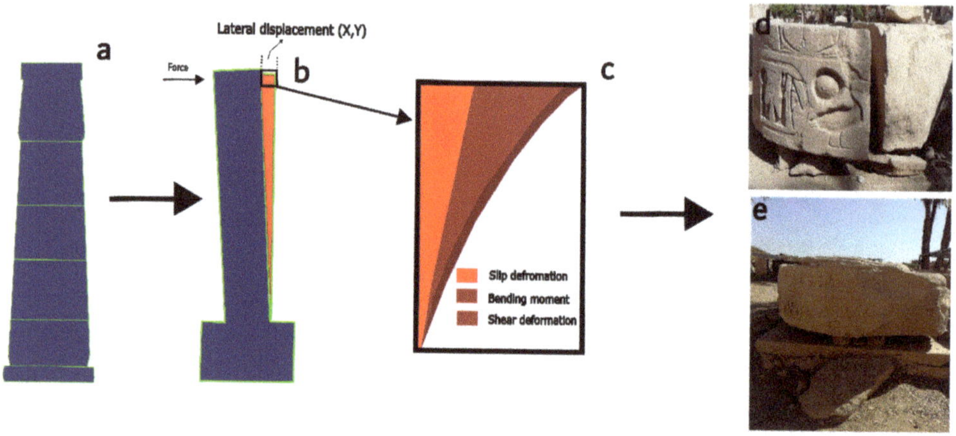

Figure 8. (**a**–**c**) Lateral displacements and deformation modes of the stone blocks in the study area in Karnak (**d**,**e**).

Figure 9. (**1**) and (**2**) Analysis of the mechanical impact of cumulative earthquakes and flooding events on the structures at Karnak. (**1a**) A general view showing scattered stone blocks, displaced due to mechanical forces, now resting on a loose red brick mastaba. (**1b**) Close-up of fractured stone blocks, illustrating the damage. (**1c**) Various stone blocks that have slid and become embedded in the soil, likely as a result of earthquake tremors or the impact of flooding waves. (**1d**–**1f**) Detailed views of mechanical chipping, fractures, and cracking observed on the stone surfaces, highlighting the extent of the damage. (**2a**) A Google Earth image of the study area, focusing on the northeastern part of Karnak, showing the current state of the stone blocks. (**2b**–**2f**) Documentation of various damage aspects of the stone blocks, including splitting, chipping, and fracturing, illustrating the ongoing deterioration of the site due to mechanical forces.

On the other hand, the flooding loads are considered a second damage factor of architectural and structural elements in the Temples of Karnak. The flooding ranged between 1 and 3 m in height and caused structural damage to the built temples. Additionally, the presence of water itself caused physical and chemical decay and alterations to stone blocks from the bottom parts of the structures. Accordingly, and hydrostatically, flooding can cause structural failures, uplifting, displacement, deformations, and the overturning of the structures. The flooding damage rate on structures can be affected by stream shape, soil composition, site topography, tree and building positions, and engineering design [66]. Wilk [67] explained the direct damage of flooding waves and surface water, as well as the groundwater flow condition and its increasing level. In this context, this is the same condition for the Temples of Karnak during flooding occurrences. Flooding events, yearly, led to an increase in the water table of the two aquifers and lakes at the site. For this, the damage to the structures and buildings was in Karnak not only from flooding water waves, height, and velocity but also from increasing levels of surface water. Furthermore, fluctuations in the water table can affect the geotechnical problems of the soil, such as changes in soil properties, swelling, shrinkage, and soil settlements [67], and this is what could cause structural damage to the temples and their structures due to the bearing problematic silty clay soil (Figure 9(1c)). Finally, the summarized causes, damage, and decay types are present in Table 2. The table provides a detailed overview of how mechanical forces, such as those from earthquakes and flooding, have impacted the structural integrity of the temple complex.

Table 2. Summary of the types of damage and decay observed in the Temples of Karnak attributed to mechanical loads. This table categorizes various forms of deterioration, including specific damage patterns, such as fractures, chipping, disintegration, and displacement of stone blocks.

Damage Type	Cause	Description	Effects on Structural Elements
Column Dismantling	Earthquakes and Flooding	Columns, especially at the bottom, were damaged due to lateral loads from seismic events and flooding pressures.	Structural instability, potential collapse, and dismantling of columns. Increased shear forces lead to low efficacy under lateral dynamic loads.
Stone Block Fracturing	Earthquakes	Stone blocks were fractured due to seismic waves transmitting through the ground, leading to sharp cracks and breakage.	Out-of-plane movements caused fractures in blocks, separating them into pieces, with partial or total failure. Vertical seismic waves caused rapid up-and-down shaking, worsening the damage.
Sharp Chippings	Earthquakes	Chipping with sharp edges formed during seismic events as a result of the forces applied on stone blocks.	Deformation of the stone surfaces.
Rounded/Peeled Chippings	Weathering	Rounded edges caused by weathering actions.	Gradual degradation of stone elements over time.
Cracking and Separation	Earthquakes and Ground Conditions	Cracking and separation occurred in stone blocks, forming fractures connected or unconnected across one or more blocks.	Stone blocks became separated and dislodged due to vertical seismic movement and unstable soil foundations.
Out-of-Plane Displacement	Earthquakes	The earthquake forces caused sections of walls to shift out of the plane.	Walls sustained severe damage, with deformation, dislodging, and partial collapse.
Soil Liquefaction	Earthquakes and High Groundwater Levels	Soil lost mechanical strength during earthquakes, causing liquefaction.	Structures sank, slid, or collapsed due to weak, liquefied soil.
Embedded Stone Blocks	Earthquakes and Weak Soil	Stone blocks became embedded in the clay soil due to ground subsidence and liquefaction.	Displacement and misalignment of structural blocks within the foundation soil.

Table 2. *Cont.*

Damage Type	Cause	Description	Effects on Structural Elements
Flooding Uplift and Displacement	Flooding	Hydrostatic forces during floods caused structural uplift and displacement.	Displacement, overturning, or uplift of entire sections of the temple during floods.
Flood-Induced Deformation	Flooding	Water pressure caused deformation, particularly in low-lying parts of structures.	Structural elements at the bottom were physically and chemically altered due to water exposure.
Block Overturning and Sliding	Earthquakes and Flooding	Stone blocks slid or overturned due to dynamic seismic forces or flood water.	Loss of structural integrity as blocks were displaced or completely overturned.
Soil Shrinkage and Swelling	Fluctuating Water Table	Soil changes in response to water level fluctuations caused shrinkage and swelling.	Contributed to the settlement of the structures, weakening the foundations and worsening structural damage over time.
Groundwater Table Fluctuations	Flooding	Increased water table and groundwater flow caused by floods or high surface water levels.	Damage due to changes in soil properties, such as settlement, swelling, and shrinkage. Structural foundations became unstable, leading to long-term deterioration.

8.2. Stone Decay, Alteration, and Water Infiltration

Stone blocks were subjected to various environmental factors, such as temperature fluctuations, wind erosion, rain, etc. In addition, the quality of stone blocks as building materials and their mineralogical content affected their durability as well. Luxor has a clear variation in the temperature from summer to winter. In summer the temperature degrees can reach up to 44 °C and the average temperature is 33 °C. The high temperature in the summer can be a thermal stress/load in the decay and degradation of the sandstone blocks. Accordingly, Shen et al. [68] explained experimentally that sandstone can be subjected to microstructure decay, such as minerals cracking and fracturing during the heating process. Additionally, in the heating period, the quartz mineral particles of sandstone can be changed volumetrically and this can cause microcracks due to the thermal stresses. Many thermal cycles and the accumulative impact of thermal loads can cause disintegration, sugaring, and flaking for the sandstone due to the stress strain actions (Figure 10a–d). Mineralogically, sandstone blocks in Karnak consist of quartz, rock fragments, clay minerals (Kaolinite and clinochlore), opaque matter, feldspar (microcline and albite), mica (muscovite), and heavy minerals. In this sense, the decay and textural features can be changed due to the sandstone composition. For this, Kompaníková et al. [69] demonstrated that sandstone that contains quartz, albite, orthoclase, muscovite, illite, and calcite is much more stable thermally than sandstone that contains chlorite, biotite, dolomite, and kaolinite and concluded that clay minerals and phyllosilicate are the most sensitive in high temperatures and can be structurally disintegrated, affecting the general fabric and color of sandstone. Wind abrasion impact was observed. Many erosion marks on the surface were identified and the erosion led to the distortion of the inscriptions (Figure 10d). Additionally, the wind caused total smoothing to the stone surfaces (Figure 10d). In this regard, after chemical alteration and disintegration due to the cycles of heating and cooling daily, weekly, and annually, wind force can weaken the cementing particles and blow them away [70]. Saltation and deflation are two processes that affect the resistance of the stone block. The main driver of stone saltation is sand and the main driver of stone deflation is the wind shear. During the saltation and deflation processes, the soil around the stone blocks can be removed and migrated to the surface of the stone to form the soiling layer (Figure 10c). Additionally, rain and strong winds can cause significant erosion, leading to the breaking and notching of the blocks. While rainfall typically does not exceed 4.5

mm, it acts as a catalyst for chemical weathering, reacting with environmental pollutants to accelerate the decay process. Accordingly, ref. [71] confirmed that smoothening, rounding, scouring at the base, pitting, and flaking are the decay patterns on wind and sand loads (Figure 10e,f).

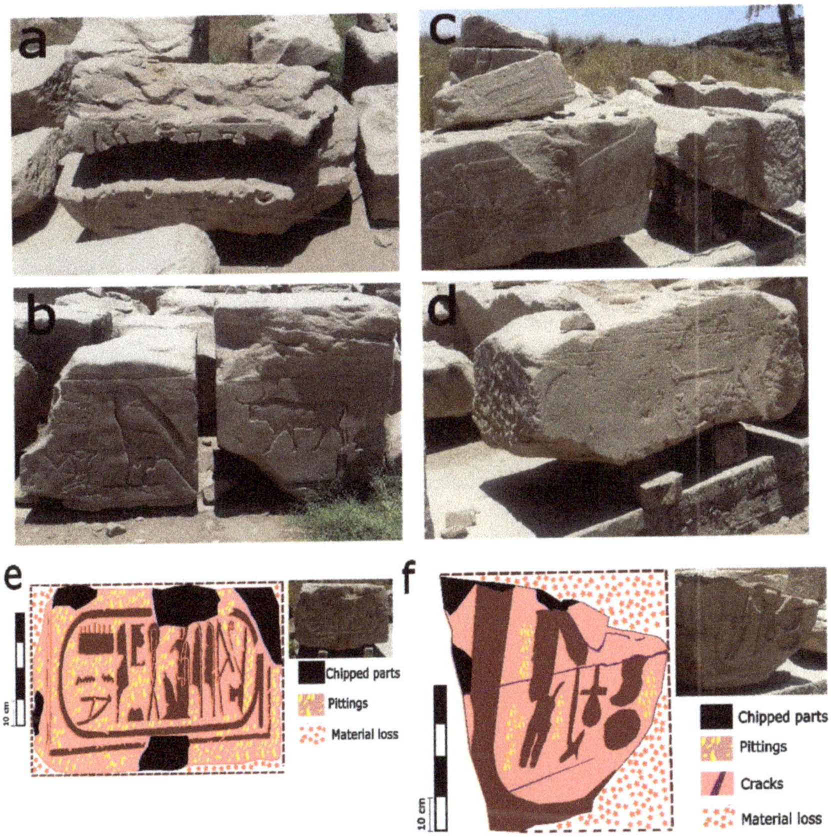

Figure 10. (**a–d**) Representation of various surface decay patterns observed on stone blocks within the Karnak Temples, including disintegration, flaking, polishing, pitting, and soiling. These patterns illustrate the effects of environmental factors on the stone material. (**e**,**f**) Detailed mapping of two specific deteriorated stone blocks, showcasing the extent and types of weathering forms present. This mapping highlights the impact of environmental conditions on the structural integrity of the stone blocks.

Contaminated water infiltration is an important factor in the alteration and degradation. Various sources of contaminated water are affecting the Temples of Karnak, such as water sources from the surrounding urbanization, agricultural activities, aquifers, canals, sacred lakes, the Nile River, and rain. Water sources were elevated and linked to the Temples of Karnak. The elevations were carried out from the east towards the Nile River and from the west towards the urban/rural and cultivated area around the archaeological site (Figure 11a–c). Contaminated components from various sources around the Temples of Karnak area were proven and studied [51,53]. Salts, such as sodium chloride (NaCl) and potassium sulfate (K_2SO_4), are the most common contaminants that were detected causing the salt weathering to the stone blocks. The salt ions can immigrate through the soil to the blocks, leaving the salts on the surface (subefflorescence) or beneath the stone surface (efflorescence) [46,72]. In the same context, salt weathering and attack on stone blocks are subjected to sequential steps: (1) transportation, (2) accumulation, (3) solute concentration, (4) precipitation, (5) fractionation, (6) local concentration, (7) local salt creeping in case of the capillarity [73]. NaCl-soluble salt crystals grow up in the largest pores and then in the smaller pores and the pressure of the crystal growth in the small pores is much more powerful [74]. Consequently, a lack of mechanical strength, breaking, and disintegration can be observed on stone surfaces (Figure 11d,e). On the other hand, clay minerals inside the sandstone of the blocks can be a serious inner defect with water presence. The chemical and petrographic analysis by [32] for four samples from various siliceous sandstone quarries as stone block sources confirmed presence of clinochlore ($(Mg,Fe^{2+},Al)_3[(OH)_2 \mid AlSi_3O_{10}]\cdot(Mg,Fe^{2+},Al)_3(OH)_6$) and kaolinite ($Al_2Si_2O_5(OH)_4$) as clay minerals. Accordingly, clinochlore is a phyllosilicate mineral (trioctahedral) and one of its characteristics is the exfoliation phenomena, mechanically [75]. However, chlorites are considered non-swelling clays; it is believed that mixed clay minerals can be considered swelling clay [76]. Clinochlore and kaolinite could be a swelling mixed clay and affected the durability of the blocks, causing flaking and exfoliation decay patterns in the presence of water (Figure 11c,e). Finally, Table 3 displays the summarized various decay and damage patterns regarding environmental and stone quality factors.

Figure 11. (**a**) Location of the case study area within Karnak, showing its elevation and relative position with respect to contaminated sources from east to west, based on global digital elevation models. (**b**) Elevation profile from west to east. (**c**) Sacred lake within the Temples of Karnak, with a yellow rectangle indicating the fluctuating water levels, which have led to the deposition of precipitates and soiling on the stone surfaces. (**d**,**e**) Examples of block surface disintegration and salt damage resulting from water impact.

Table 3. Different types of damage and decay affecting sandstone blocks in Karnak, based on the environmental factors and mineralogical content.

Type of Damage/Decay	Cause	Description/Effect
Thermal (Stress/Load)	Temperature fluctuations	Microstructure decay (mineral cracking and fracturing), volumetric changes in quartz leading to microcracks, sugaring, flaking, and disintegration.
Mineralogical Decay	Composition of sandstone (quartz, albite, muscovite, etc.)	Sandstone with chlorite, biotite, dolomite, and kaolinite is less thermally stable. Clay minerals (e.g., kaolinite, illite) and phyllosilicates disintegrate at high temperatures.
Wind (Erosion/Abrasion)	Wind-driven sand and particles	Surface erosion, distortion of inscriptions, total polishing, rounding, scouring, pitting, flaking, and weakening of cementing particles.
Salt Weathering	Salts from contaminated water sources	Salt ions (NaCl, K_2SO_4) migrate through the soil and block the surface, leading to subefflorescence and efflorescence. Crystal growth in pores causes internal stresses and disintegration.
Capillary Action (Salt Creep)	Water movement through pores	Damage is caused through transportation, accumulation, solute concentration, precipitation, crystal fractionation, and the creeping of salts causing internal and surface damage.
Contaminated Water Infiltration	Water sources (urbanization, agriculture, canals, Nile, rain)	Waterborne contaminants lead to salt deposition, mechanical weakening, and breakage of the stone blocks.
Clay Swelling and Exfoliation	Presence of swelling clays (clinochlore, kaolinite)	Swelling of clay minerals in sandstone leads to inner defects, causing flaking and exfoliation, especially in the presence of water.

8.3. Vegetation Change Detection

Copernicus Global (NDVI raster 300 V2) was used for the observation and detection of the vegetation change using mid to low spatial resolution. NDVI is an indicator to quantify the vegetation's greenness and density. Measurement of NDVI can be calculated through Formula (1):

$$\text{NDVI} = (\text{REF\_nir} - \text{REF\_red})/(\text{REF\_nir} + \text{REF\_red}) \quad (1)$$

where REF_nir and REF_red are the spectral reflectances measured in the near infrared and red wavelengths, respectively.

Figure 12a,b and Table 4 show the multi-temporal analysis and monitoring of the vegetation index and biomass in the Luxor city and case study area (in and around Karnak) from August 2023 to February 2024. For vegetation coverage and density, NDVI values range from −1.0 to 1.0. Negative values indicate water presence and positive values near zero refer to bare soil. In addition, values from 0.1 to 0.5 refer to sparse vegetation and values from 0.6 and up are referring to dense green vegetation. Accordingly, the results (Figure 12b) illustrated the observed variation of vegetation coverage temporally from bare to dense vegetation, which confirms the activity of water resources and its fluctuations in the Karnak area. For instance, in August and September 2023, the vegetation coverage fluctuation was slightly observed. The values were slightly increasing in the last ten days of August and September and some parts were densely covered with vegetation. In January and February, variations of vegetation mass were observed, especially between the last ten days in January (spare) and the first ten days (dense). Figure 12c,d shows the vegetation and plants around the stone blocks, as well as the biophysical and biochemical impact of the vegetation and plants on the stone blocks' resilience.

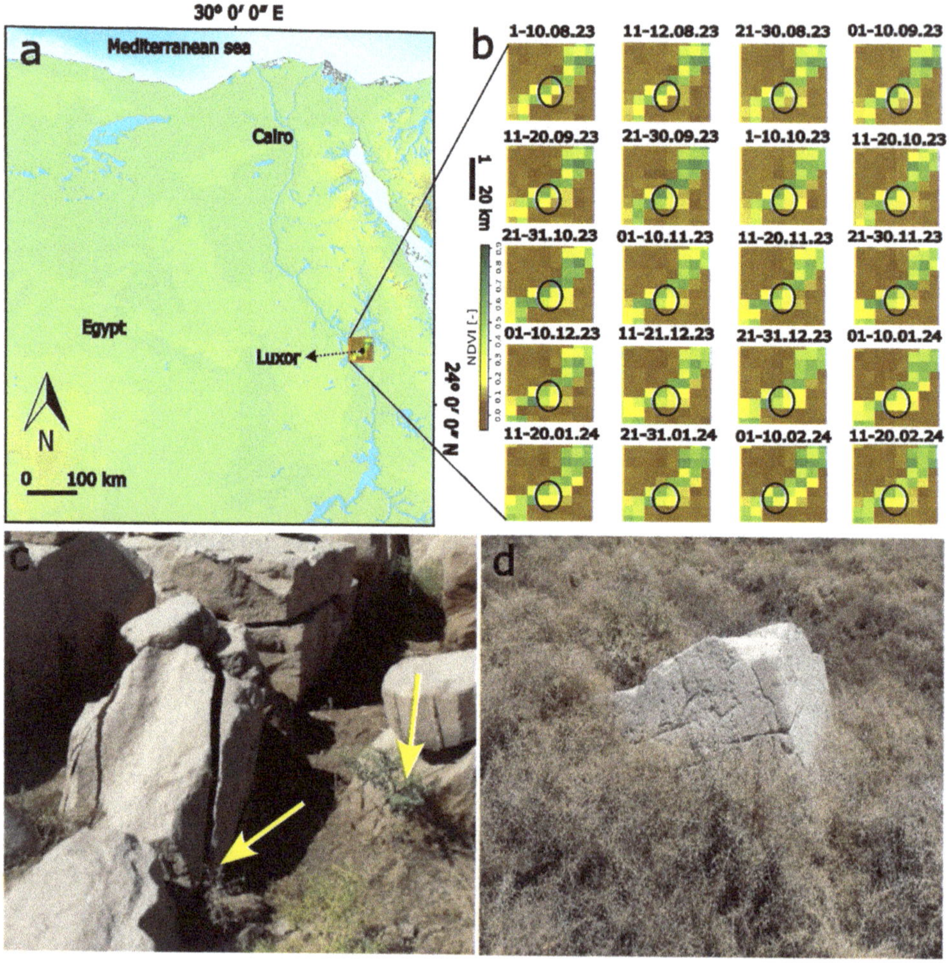

Figure 12. (**a**) General map of Egypt indicating the Luxor area with the NDVI Index. (**b**) Multi-temporal monitoring for the vegetation variations in Luxor and the case study area using the NDVI Index. (**c**,**d**) Loss of mechanical strength and physical and aesthetical properties due to the biodeterioration of the stone blocks and shown by yellow arrows.

Table 4. Temporal variation of vegetation coverage and density in Luxor city, particularly in the Karnak area, based on the NDVI values.

Month/Period	NDVI Range (Coverage)	Vegetation Coverage	NDVI Range (Density)	Vegetation Density
August 2023 (early)	Near 0 to 0.1	Bare soil/sparse vegetation	0.1–0.5	Sparse vegetation
August 2023 (late)	0.1–0.5	Sparse vegetation	0.5–0.6	Moderately dense vegetation
September 2023	0.1–0.5	Sparse vegetation	0.5–0.6	Increasingly dense vegetation
January 2024 (late)	Near 0 to 0.2	Bare soil/sparse vegetation	0.1–0.3	Sparse vegetation
February 2024 (early)	0.5–0.7	Dense vegetation	0.6 and up	Dense green vegetation

Plants with spinous structures, such as *Atraphaxis* and thorns/spines/prickles, were found to be superior plants that cause biodegradation to the stone blocks. The physical and morphological features of the plants indicated their impact on the mechanical damage of the stone and caused separation and fracturing of stone blocks (Figure 12c). In this regard, the biodeterioration on stones can be classified into physical/mechanical, chemical, and aesthetic [77]. Mechanically, the growth and movements of plants and organisms cause irreversible deformations under permanent loads. Chemically, some organisms and plants use the internal components of the stones as food materials (assimilation). In addition, plants and organisms can produce substances (dissimilatory), such as acids, that lead to chemical alteration and etching to the stone surfaces. Furthermore, plants and organisms can affect the aesthetical characteristics of stone due to the metabolism and byproducts causing encrustation and biofilms.

9. Restoration and Conservation Plan for Sandstone Blocks at Karnak

The sandstone blocks at Karnak have been subjected to various environmental factors, such as temperature fluctuations, wind erosion, salt weathering, contaminated water infiltration, and mechanical loads. To effectively conserve and restore the blocks, a comprehensive plan is required, with a focus on constructing new mastabas (elevated platforms) to prevent soil and water contact and to create an open museum setting for the Karnak site. Accordingly, for suitable restoration and conservation strategy, the next steps can be implemented, such as (1) constructing the mastabas (Elevated Platforms); (2) addressing decay mechanisms; (3) water management and contamination prevention; (4) conservation of affected blocks; (5) establishing an open museum; and (6) long-term preservation strategies.

The primary aim is to lift the sandstone blocks off the ground, minimizing direct contact with soil and water sources, thus preventing further water-related degradation, such as salt weathering and clay swelling. The mastabas (Figure 13a–d) should be constructed from durable, non-porous materials, like granite. In addition, each platform should be raised high enough to prevent capillary action from drawing contaminated water into the blocks, around 40–60 cm (Figure 13b). Drainage systems can be integrated around the mastabas to direct water away from the blocks. Survey and assess the condition of each sandstone block to determine specific restoration needs before relocation and use appropriate lifting and handling techniques to avoid further structural damage during the process of raising the blocks onto the mastabas. A protective layer (e.g., waterproof membranes) can be applied between the block and the platform to ensure complete isolation from moisture and water infiltration. Additionally, implement air circulation systems to avoid moisture buildup under the blocks and future decay. For restoration of the blocks heavily affected by surface erosion or salt weathering, mechanical and chemical cleaning methods should be employed carefully to restore the blocks. Using non-invasive cleaning techniques, such as micro-sandblasting or laser cleaning, for removing salt crusts, stains, and surface contaminants is considered a good option in case of their availability. Furthermore, consolidation treatments using ethyl silicate-based or lime-based solutions strengthen weak or fragmented stone areas and prevent further disintegration. In areas where the stone blocks have suffered structural damage, careful repair using compatible materials should be undertaken. Cracks can be filled with lime mortar.

For management and sustainable development, the relocated and restored sandstone blocks can be displayed as part of an open museum in the Karnak Temples. Each block should have informational signage detailing its original location, historical significance, and the restoration process it underwent. Moreover, designated pathways for visitors should be constructed to ensure easy access while minimizing direct human interaction with the blocks. These paths should be made from non-invasive materials that will not contribute to further contamination. Regular monitoring of the blocks' condition should be carried out using techniques like 3D scanning to track any signs of decay or deterioration.

Finally, a preventive conservation schedule should be established for cleaning, desalination, and reapplying protective coatings as necessary.

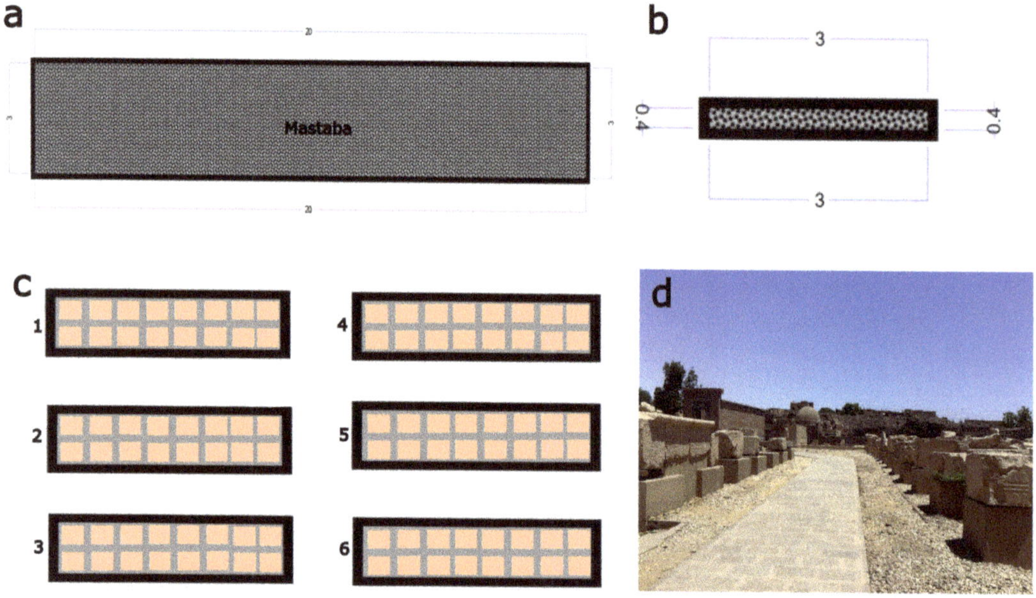

Figure 13. (**a**) The dimension of a Mastaba (20 m × 3 m); (**b**) cross-section of a Mastaba and its dimensions (3 m × 40 cm); (**c**) suggested system for six Mastabas constructions to lift around 144 stone blocks in the northeast corner (phase 1) over them. (**d**) Conducted a previous project to lift the stone blocks upon the Mastabas as part of the Karnak Open Museum (in front of the chapel of Hatshepsut), executed by the Egyptian Ministry of Tourism and Antiquity.

10. Conclusions

This detailed study of the stone blocks in the northeast corner of the Karnak Temples has exposed a range of sustainability challenges attributed to various mechanical and geochemical damage factors. In situ assessments reveal that these stone blocks are suffering from substantial structural degradation, including dismantling fractures and scattering. This damage is primarily the result of seismic activities, flooding events, and anthropogenic interference. The assessment also underscores the detrimental effects of contaminated water sources, adverse climatic conditions, and invasive vegetation on the durability of the stone. Multi-temporal satellite imagery has proven invaluable in tracking changes in vegetation coverage, which, in turn, influences the degradation process. These findings highlight the urgent need for ongoing monitoring, comprehensive documentation, and the application of advanced conservation techniques to ensure the preservation of the Karnak Temples.

Looking forward, future research should focus on several key areas to enhance the understanding and management of these challenges. This includes the application of non-destructive testing (NDT) methods to assess the condition of the materials without causing further damage. Detailed classification and analysis of the materials will be crucial for developing tailored conservation strategies. Additionally, the formulation of extensive restoration and site management strategies is essential for addressing the multi-faceted issues identified. Such strategies should incorporate robust conservation methods designed to protect the historical and cultural significance of the Karnak Temples for future generations. Interdisciplinary collaboration will be critical in this endeavor, integrating expertise from archaeology, geology, and conservation to tackle the complex challenges associated with preserving this ancient heritage site.

To effectively manage and mitigate the identified challenges, a comprehensive conservation and restoration strategy must be implemented. Key components of this strategy include the construction of elevated platforms to reduce soil and water contact with the stone blocks. These platforms should be designed from durable materials, like granite, and raised sufficiently (40–60 cm) to prevent capillary action and water infiltration. Integrated drainage systems around these platforms will help direct water away from the blocks, further protecting them from moisture-related damage.

Addressing the decay mechanisms involves employing non-invasive cleaning techniques, such as micro-sandblasting or laser cleaning, to remove salt crusts and surface contaminants. Consolidation treatments using ethyl silicate-based or lime-based solutions will strengthen weakened areas of the stone. Effective water management and contamination prevention measures are also crucial. Protective layers, such as waterproof membranes, should be applied between the blocks and platforms to prevent moisture infiltration and air circulation systems can mitigate moisture buildup under the blocks.

For blocks suffering from severe erosion or salt weathering, careful restoration using compatible materials is necessary. Cracks should be filled with lime mortar and structural repairs should be conducted with a focus on maintaining historical accuracy. The creation of an open museum setting to display the relocated and restored blocks will enhance public understanding and appreciation of the site. Informational signage detailing the blocks' original locations, historical significance, and restoration processes will further educate visitors.

Long-term preservation efforts should include regular monitoring using advanced techniques, such as 3D scanning, to track the condition of the blocks and detect early signs of deterioration. A preventive conservation schedule should be established, incorporating periodic cleaning, desalination, and reapplication of protective coatings to ensure ongoing protection of the stone blocks.

Author Contributions: Conceptualization, A.F., E.M.-P. and S.D.-B.; writing—original draft preparation, A.F.; writing—review and editing, A.F.; visualization, A.F.; supervision, E.M.-P. and S.D.-B. All authors have read and agreed to the published version of the manuscript.

Funding: This research received no external funding.

Data Availability Statement: Data sharing is not applicable to this article, as no datasets were generated or analysed during the current study.

Conflicts of Interest: The authors declare no conflict of interest.

References

1. Wang, F.; Frühwirt, T.; Konietzky, H. Influence of repeated heating on physical-mechanical properties and damage evolution of granite. *Int. J. Rock Mech. Min. Sci.* **2020**, *136*, 104514. [CrossRef]
2. Brimblecombe, P. Environment and architectural stone. In *Springer eBooks*; Springer: Berlin/Heidelberg, Germany, 2014; pp. 317–347. [CrossRef]
3. Brimblecombe, P. Heritage Climatology. In *Climate Change and Cultural Heritage: Proceedings of the Ravello International Workshop, 14–16 May 2009 and Strasbourg European Master-Doctorate Course, 7–11 September 2009*; Edipuglia: Bari, Italy, 2010; pp. 54–57. Available online: https://www.getty.edu/conservation/publications_resources/teaching/pdf/stone_conservation/module4/7_SO.Brimblecombe.14May13a.pdf (accessed on 7 July 2024).
4. Ito, W.H.; Scussiato, T.; Vagnon, F.; Ferrero, A.M.; Migliazza, M.R.; Ramis, J.; De Queiroz, P.I.B. On the Thermal Stresses due to Weathering in Natural Stones. *Appl. Sci.* **2021**, *11*, 1188. [CrossRef]
5. Founda, D.; Katavoutas, G.; Pierros, F.; Mihalopoulos, N. Centennial changes in heat waves characteristics in Athens (Greece) from multiple definitions based on climatic and bioclimatic indices. *Glob. Planet. Change* **2022**, *212*, 103807. [CrossRef]
6. Livada, I.; Santamouris, M.; Niachou, K.; Papanikolaou, N.; Mihalakakou, G. Determination of places in the great Athens area where the heat island effect is observed. *Theor. Appl. Climatol.* **2002**, *71*, 219–230. [CrossRef]
7. Katavoutas, G.; Founda, D.; Kitsara, G.; Giannakopoulos, C. Climate change and thermal comfort in top tourist Destinations—The case of Santorini (Greece). *Sustainability* **2021**, *13*, 9107. [CrossRef]

8. Shi, X.J.; Shi, X.F. Numerical prediction on erosion damage caused by wind-blown sand movement. *Eur. J. Environ. Civ. Eng.* **2014**, *18*, 550–566. [CrossRef]
9. Belfiore, C.; Calabrò, C.; Ruffolo, S.; Ricca, M.; Török, Á.; Pezzino, A.; La Russa, M. The susceptibility to degradation of stone materials used in the built heritage of the Ortygia island (Syracuse, Italy): A laboratory study. *Int. J. Rock Mech. Min. Sci.* **2021**, *146*, 104877. [CrossRef]
10. Striani, R.; Cappai, M.; Casnedi, L.; Corcione, C.E.; Pia, G. Coating's influence on wind erosion of porous stones used in the Cultural Heritage of Southern Italy: Surface characterisation and resistance. *Case Stud. Constr. Mater.* **2022**, *17*, e01501. [CrossRef]
11. Fahmy, A.; Domínguez-Bella, S.; Martínez-López, J.; Molina-Piernas, E. Sand dune movement and flooding risk analysis for the pyramids of Meroe, Al Bagrawiya archaeological site, Sudan. *Herit. Sci.* **2023**, *11*, 136. [CrossRef]
12. Hemeda, S.; Fahmy, A.; Sonbol, A. Geo-Environmental and structural problems of the first successful true pyramid, (Snefru Northern Pyramid) in Dahshur, Egypt. *Geotech. Geol. Eng.* **2018**, *37*, 2463–2484. [CrossRef]
13. The Getty Conservation Institute. Archaeological Sites: Conservation and management. In *Readings in Conservation*; Sullivan, S., Mackay, R., Eds.; Getty Conservation Institute: Los Angeles, CA, USA, 2012; Available online: https://www.getty.edu/publications/resources/virtuallibrary/9781606061244.pdf (accessed on 12 July 2024).
14. Reader, C.D. A Geomorphological Study of the Giza Necropolis, with Implications for the Development of the Site. *Archaeometry* **2001**, *43*, 149–165. [CrossRef]
15. Faranda, D.; Ginesta, M.; Alberti, T.; Coppola, E.; Anzidei, M. Attributing Venice Acqua Alta events to a changing climate and evaluating the efficacy of MoSE adaptation strategy. *Npj Clim. Atmos. Sci.* **2023**, *6*, 181. [CrossRef]
16. Carbognin, L.; Teatini, P.; Tomasin, A.; Tosi, L. Global change and relative sea level rise at Venice: What impact in term of flooding. *Clim. Dyn.* **2009**, *35*, 1039–1047. [CrossRef]
17. Oliveira, C.S. Seismic Vulnerability of Historical Constructions: A Contribution. *Bull. Earthq. Eng.* **2003**, *1*, 37–82. [CrossRef]
18. Kázmér, M. Damage to Ancient Buildings from Earthquakes. In *Springer eBooks*; Springer: Berlin/Heidelberg, Germany, 2015; pp. 500–506. [CrossRef]
19. Park, H.; Ha, J.; Kim, S.; Jo, S. Seismic performance of ancient masonry structures in Korea rediscovered in 2016 M 5.8 Gyeongju earthquake. *Sustainability* **2019**, *11*, 1565. [CrossRef]
20. Marco, S. Recognition of earthquake-related damage in archaeological sites: Examples from the Dead Sea fault zone. *Tectonophysics* **2008**, *453*, 148–156. [CrossRef]
21. Sallam, M.A.; Hassan, H.M.; Sayed, M.A.; Hafiez HE, A.; Zahra, H.S.; Salem, M. Seismic vulnerability assessment of historical minarets in Cairo. *Geoenviron. Disasters* **2023**, *10*, 30. [CrossRef]
22. Bottari, C.; Capizzi, P.; Sortino, F. Unraveling the Seismic Source in Archaeoseismology: A combined approach on local site effects and geochemical data integration. *Heritage* **2024**, *7*, 427–447. [CrossRef]
23. Kreibich, H.; Thieken, A.H.; Grunenberg, H.; Ullrich, K.; Sommer, T. Extent, perception and mitigation of damage due to high groundwater levels in the city of Dresden, Germany. *Nat. Hazards Earth Syst. Sci.* **2009**, *9*, 1247–1258. [CrossRef]
24. Bork, M.; Lange, J.; Graf-Rosenfellner, M.; Hensen, B.; Olsson, O.; Hartung, T.; Fernández-Pascual, E.; Lang, F. Urban storm water infiltration systems are not reliable sinks for biocides: Evidence from column experiments. *Sci. Rep.* **2021**, *11*, 7242. [CrossRef]
25. Delgado, J.M.P.Q.; Guimarães, A.S.; De Freitas, V.P.; Antepara, I.; Kočí, V.; Černý, R. Salt damage and rising damp treatment in building structures. *Adv. Mater. Sci. Eng.* **2016**, *2016*, 1280894. [CrossRef]
26. Lasaponara, R.; Murgante, B.; Elfadaly, A.; Qelichi, M.; Shahraki, S.; Wafa, O.; Attia, W. Spatial Open Data for Monitoring Risks and Preserving Archaeological Areas and Landscape: Case Studies at Kom el Shoqafa, Egypt and Shush, Iran. *Sustainability* **2017**, *9*, 572. [CrossRef]
27. Fahmy, A.; Gołąbiewska, A.; Wojnicz, W.; Stanisławska, A.; Kowalski, J.; Łuczak, J.; Zaleska-Medynska, A.; Domínguez-Bella, S.; Martínez-López, J.; Molina-Piernas, E. Multi-functional monodispersed SiO_2–TiO_2 core-shell nanostructure and TEOS in the consolidation of archaeological lime mortars surfaces. *J. Build. Eng.* **2023**, *79*, 107809. [CrossRef]
28. Fouad, S.S.; Heggy, E.; Weilacher, U. Waterways transformation in the vulnerable port city of Alexandria. *Cities* **2023**, *141*, 104426. [CrossRef]
29. Oguchi, C.T.; Yu, S. A review of theoretical salt weathering studies for stone heritage. *Prog. Earth Planet. Sci.* **2021**, *8*, 32. [CrossRef]
30. Lubelli, B.; Van Hees, R.P.; Groot, C.J. The role of sea salts in the occurrence of different damage mechanisms and decay patterns on brick masonry. *Constr. Build. Mater.* **2004**, *18*, 119–124. [CrossRef]
31. Rombolà, R.M. The GIS for Urban Analysis and Risk Map of Archaeological Site: The Case of Sphinxes Avenue in Luxor. In *Sustainable Conservation and Urban Regeneration. Research for Development*; Folli, M., Ed.; Springer: Cham, Switzerland, 2018. [CrossRef]
32. Fitzner, B.; Heinrichs, K.; la Bouchardiere, D. Weathering damage on Pharaonic sandstone monuments in Luxor-Egypt. *Build. Environ.* **2003**, *38*, 1089–1103. [CrossRef]
33. Ahmed, A.A. Land use change and deterioration of pharaonic monuments in upper Egypt. *J. Eng. Sci. Assiut Univ.* **2009**, *37*, 161. Available online: https://jesaun.journals.ekb.eg/article_121202_46d5400d2a495a6626e99b8e8331681d.pdf (accessed on 10 July 2024). [CrossRef]
34. Megahed, H.A. Hydrological and archaeological studies to detect the deterioration of Edfu temple in Upper Egypt due to environmental changes during the last five decades. *SN Appl. Sci.* **2020**, *2*, 1952. [CrossRef]

35. Marey Mahmoud, H.; Kantiranis, N.; Stratis, J. Salt damage on the wall paintings of the festival temple of Thutmosis III, Karnak Temples Comlpex, Upper of Egypt. A case study. *Int. J. Commun. Syst.* **2010**, *1*, 133–142.
36. Fahmy, A.; Martínez-López, J.; Sánchez-Bellón, Á.; Domínguez-Bella, S.; Molina-Piernas, E. Multianalytical diagnostic approaches for the assessment of materials and decay of the archaeological sandstone of Osiris Temple (The Abaton) in Bigeh Island, Philae (Aswan, Egypt). *J. Cult. Herit.* **2022**, *58*, 167–178. [CrossRef]
37. Fahmy, A.; Molina-Piernas, E.; Martínez-López, J.; Domínguez-Bella, S. Salt weathering impact on Nero/Ramses II Temple at El-Ashmonein archaeological site (Hermopolis Magna), Egypt. *Herit. Sci.* **2022**, *10*, 125. [CrossRef]
38. Wilkinson, R.H. *The Complete Temples of Ancient Egypt*; Thames & Hudson: London, UK, 2000. [CrossRef]
39. Dieter, A. *The Encyclopedia of Ancient Egyptian Architecture*; Tauris, I.B., Ed.; Princeton University Press: Princeton, NJ, USA, 2002. Available online: https://archive.org/search.php?query=external-identifier:%22urn:oclc:record:1033591802%22 (accessed on 11 July 2024).
40. Elaine, S. Introduction. In *Digital Karnak, Los Angeles*; University of California Santa Cruz: Santa Cruz, CA, USA, 2008. Available online: http://dlib.etc.ucla.edu/projects/Karnak (accessed on 11 July 2024).
41. Blyth, E. *Karnak: Evolution of a Temple*; Routledge: London, UK, 2006; ISBN 0-415-40486-X. Available online: https://api.pageplace.de/preview/DT0400.9781134136681_A24932187/preview-9781134136681_A24932187.pdf (accessed on 15 July 2024).
42. Ghilardi, M.; Boraik, M. Reconstructing the holocene depositional environments in the western part of Ancient Karnak temples complex (Egypt): A geoarchaeological approach. *J. Archaeol. Sci.* **2011**, *38*, 3204–3216. [CrossRef]
43. Ahmed, A.A.; Fogg, G.E.; Gameh, M.A. Water use at Luxor, Egypt: Consumption analysis and future demand forecasting. *Environ. Earth Sci.* **2014**, *72*, 1041–1053. [CrossRef]
44. Elkhateeb, S.O.; Dosoky, W.; Mohamed, A. Geoelectric and mineralogical studies for foundation soil characterization in new Luxor city, Upper Egypt. *Arab. J. Geosci.* **2022**, *15*, 1220. [CrossRef]
45. Harrell, J.A.; Storemyr, P. Ancient Egyptian quarries—An illustrated overview. In *QuarryScapes: Ancient Stone Quarry Landscapes in the Eastern Mediterranean*; Abu-Jaber, N., Bloxam, E.G., Degryse, P., Heldal, T., Eds.; Geological Survey of Norway Special Publication: Fribourg, Switzerland, 2009; Volume 12, pp. 7–50. Available online: https://www.ngu.no/upload/Publikasjoner/Special%20publication/SP12_s7-50.pdf (accessed on 17 July 2024).
46. Fahmy, A.; Molina-Piernas, E.; Martínez-López, J.; Machev, P.; Domínguez-Bella, S. Coastal Environment Impact on the Construction Materials of Anfushi's Necropolis (Pharos's Island) in Alexandria, Egypt. *Minerals* **2022**, *12*, 1235. [CrossRef]
47. Fahmy, A.; Molina-Piernas, E.; Domínguez-Bella, S.; Martínez-López, J.; Helmi, F. Geoenvironmental investigation of Sahure's pyramid, Abusir archeological site, Giza, Egypt. *Herit. Sci.* **2022**, *10*, 61. [CrossRef]
48. Nilsson, M.; Ward, J. Gebel el-Silsila throughout the Ages: Part 7-Late Period to Grieco Roman Era. 2021. Available online: https://lucris.lub.lu.se/ws/portalfiles/portal/113150354/AE124NilssonGebel7.pdf (accessed on 18 July 2024).
49. Nilsson, M.; Ward, J. Gebel el-Silsila throughout the Ages: Part 1. 2019. Available online: https://lucris.lub.lu.se/ws/portalfiles/portal/101678193/AE113WardGebel1aspublished.pdf (accessed on 18 July 2024).
50. Nilsson, M.; Almasy, A. Quarrying for Claudius, protected by Min: A small quarry at Gebel el-Silsila East. *Br. Mus. Stud. Anc. Egypt Sudan* **2015**, *22*, 87–110. Available online: https://web.archive.org/web/20180409222029id_/http://www.britishmuseum.org/pdf/Nilsson_FINAL.pdf (accessed on 13 July 2024).
51. Elwaseif, M.; Ismail, A.; Abdalla, M.; Abdel-Rahman, M.; Hafez, M.A. Geophysical and hydrological investigations at the west bank of Nile River (Luxor, Egypt). *Environ. Earth Sci.* **2012**, *67*, 911–921. [CrossRef]
52. Mokhtar, M. Monitoring of Groundwater and Ancient Structures in Luxor and Karnak Temples. 2007. Available online: https://arce.org/ (accessed on 13 July 2024).
53. Ismail, A.; Anderson, N.L.; Rogers, J.D.; Society of Exploration Geophysicists. Hydrogeophysical investigation at Luxor, Southern Egypt. *J. Environ. Eng. Geophys.* **2005**, *10*, 35–49. [CrossRef]
54. SECAP. Sustainable Energy and Climate Action Plan. City of Luxor, Governorate of Luxor—Egypt. 2017. Available online: https://www.climamed.eu/wp-content/uploads/files/Egypt-Governorate-of-Luxor-Sustainable-Energy-and-Climate-Action-Plan-(SECAP).pdf (accessed on 19 July 2024).
55. Qadri, S.T.; Mirza, M.Q.; Raja, A.; Yaghmaei-Sabegh, S.; Hakimi, M.H.; Ali, S.H.; Khan, M.Y. Application of Probabilistic Seismic Hazard Assessment to Understand the Earthquake Hazard in Attock City, Pakistan: A Step towards Linking Hazards and Sustainability. *Sustainability* **2023**, *15*, 1023. [CrossRef]
56. Karakhanyan, A.; Armenia, Y.; Avagyan, A.; Sourouzian, H. Archaeoseismological studies at the temple of Amenhotep III, Luxor, Egypt. In *Ancient Earthquakes: Geological Society of America Special Paper*; Sin tubin, M., Stewart, I.S., Niemi, T.M., Altunel, E., Eds.; Geological Society of America: Boulder, CO, USA, 2010; p. 471. [CrossRef]
57. El-Sayed, A.; Wahlström, R.; Kulhánek, O. Seismic hazard of Egypt. *Nat. Hazards* **1994**, *10*, 247–259. Available online: https://gfzpublic.gfz-potsdam.de/rest/items/item_232445_1/component/file_232444/content (accessed on 19 July 2024). [CrossRef]
58. Toonen WH, J.; Graham, A.; Pennington, B.T.; Hunter, M.A.; Strutt, K.D.; Barker, D.S.; Masson-Berghoff, A.; Emery, V.L. Holocene fluvial history of the Nile's west bank at ancient Thebes, Luxor, Egypt, and its relation with cultural dynamics and basin-wide hydroclimatic variability. *Geoarchaeology* **2018**, *33*, 273–290. [CrossRef]
59. Russell, C. Karnak before the Spade Arrived. 2005. Available online: https://www.researchgate.net/publication/348133958_Karnak_Before_the_French_Arrived (accessed on 20 July 2024).

60. Toonen WH, J.; Graham, A.; Masson-Berghoff, A.; Peeters, J.; Winkels, T.G.; Pennington, B.T.; Hunter, M.A.; Strutt, K.D.; Barker, D.S.; Emery, V.L.; et al. Amenhotep III's Mansion of Millions of Years in Thebes (Luxor, Egypt): Submergence of high grounds by river floods and Nile sediments. *J. Archaeol. Sci. Rep.* **2019**, *25*, 195–205. [CrossRef]
61. McLane, J.; Raphael AJ Wüst Porter, B.; Rutherford, J. Flash-Flood Impacts and Protection Measures in the Valley of the Kings, Luxor, Egypt. *APT Bull. J. Preserv. Technol.* **2003**, *34*, 37–45. [CrossRef]
62. Baines, J. The inundation stela of Sebekhotpe VIII. *Acta Orient.* **1974**, *36*, 39–54. Available online: https://ora.ouls.ox.ac.uk/objects/uuid:f0914ff3-bc93-4525-951d-870f1782fe99 (accessed on 20 July 2024).
63. Rodríguez-Pascua, M.A.; Pérez-López, R.; Giner-Robles, J.L.; Garduño-Monroy, V.H. A comprehensive classification of Earthquake Archaeological Effects (EAE) in archaeoseismology: Application to ancient remains of Roman and Mesoamerican cultures. *Quat. Int.* **2011**, *242*, 20–30. [CrossRef]
64. Yön, B.; Sayin, E.; Onat, O. Earthquakes and structural damages. In *InTech eBooks*; IntechOpen: London, UK, 2017. [CrossRef]
65. Feilden, B.; Alva, A. Earthquakes and Historic Buildings. 1993. Available online: https://openarchive.icomos.org/id/eprint/843/1/ro29.pdf (accessed on 22 July 2024).
66. Hughes, R. The effects of flooding upon the buildings in developing countries. *Disasters* **1982**, *6*, 183–194. Available online: https://www.eird.org/bibliovirtual/riesgo-urbano/pdf/eng/doc12030/doc12030-a.pdf (accessed on 22 July 2024). [CrossRef]
67. Wilk, K. Hazards for buildings and structures caused by flood conditions. *E3S Web Conf.* **2018**, *45*, 00101. [CrossRef]
68. Shen, Y.-J.; Zhang, Y.-L.; Gao, F.; Yang, G.-S.; Lai, X.-P. Influence of Temperature on the Microstructure Deterioration of Sandstone. *Energies* **2018**, *11*, 1753. [CrossRef]
69. Kompaníková, Z.; Gomez-Heras, M.; Michňová, J.; Durmeková, T.; Vlčko, J. Sandstone alterations triggered by fire-related temperatures. *Environ. Earth Sci.* **2014**, *72*, 2569–2581. [CrossRef]
70. Liang, Z.; Wu, Z.; Yao, W.; Noori, M.; Yang, C.; Xiao, P.; Leng, Y.; Deng, L. Pisha sandstone: Causes, processes and erosion options for its control and prospects. *Int. Soil Water Conserv. Res.* **2019**, *7*, 1–8. [CrossRef]
71. Richards, J.; Viles, H.; Guo, Q. The importance of wind as a driver of earthen heritage deterioration in dryland environments. *Geomorphology* **2020**, *369*, 107363. [CrossRef]
72. Alves, C.; Figueiredo, C.A.M.; Sanjurjo-Sánchez, J.; Hernández, A.C. Salt Weathering of Natural Stone: A Review of Comparative Laboratory Studies. *Heritage* **2021**, *4*, 1554–1565. [CrossRef]
73. Zehnder, K.; Arnold, A. Crystal growth in salt efflorescence. *J. Cryst. Growth* **1989**, *97*, 513–521. [CrossRef]
74. Cardell, C.; Delalieux, F.; Roumpopoulos, K.; Moropoulou, A.; Auger, F.; van Grieken, R. Salt-induced decay in calcareous stone monuments and buildings in a marine environment in SW France. In *Construction and Building Materials*; Elsevier Ltd.: Amsterdam, The Netherlands, 2003; Volume 17. [CrossRef]
75. De Oliveira, R.; Guallichico, L.A.G.; Policarpo, E.; Cadore, A.R.; Freitas, R.O.; da Silva, F.M.C.; de CTeixeira, V.; Paniago, R.M.; Chacham, H.; Matos, M.J.S.; et al. High throughput investigation of an emergent and naturally abundant 2D material: Clinochlore. *Appl. Surf. Sci.* **2022**, *599*, 153959. [CrossRef]
76. Elert, K.; Rodriguez-Navarro, C. Degradation and conservation of clay-containing stone: A review. In *Construction and Building Materials*; Elsevier Ltd.: Amsterdam, The Netherlands, 2022; Volume 330. [CrossRef]
77. Cozzolino, A.; Adamo, P.; Bonanomi, G.; Motti, R. The Role of Lichens, Mosses, and Vascular Plants in the Biodeterioration of Historic Buildings: A Review. *Plants* **2022**, *11*, 3429. [CrossRef]

minerals

MDPI

Article

Materials and Techniques of the Mural Paintings in the Church-Ossuary of the Rila Monastery, Bulgaria

Bistra Stamboliyska [1,*], Stefan Tapanov [2], Evelina Velcheva [1], Stela Atanasova-Vladimirova [3], Bogdan Ranguelov [3], Maya Guncheva [1], Simeon Stoyanov [1] and Denitsa Yancheva [1,4]

1 Institute of Organic Chemistry with Centre of Phytochemistry, Bulgarian Academy of Sciences, Acad. G. Bonchev Str., Build. 9, 1113 Sofia, Bulgaria; evelina.velcheva@orgchm.bas.bg (E.V.); maya.guncheva@orgchm.bas.bg (M.G.); simeon.stoyanov@orgchm.bas.bg (S.S.); denitsa.pantaleeva@orgchm.bas.bg (D.Y.)

2 National Academy of Art, Faculty of Applied Arts, 1113 Sofia, Bulgaria; tapanow.stefan@gmail.com

3 Institute of Physical Chemistry, Bulgarian Academy of Sciences, Acad. G. Bonchev Str., Build. 11, 1113 Sofia, Bulgaria; statanasova@ipc.bas.bg (S.A.-V.); rangelov@ipc.bas.bg (B.R.)

4 Department of Organic Chemistry, University of Chemical Technology and Metallurgy, 8 Kliment Ohridski Blvd., 1756 Sofia, Bulgaria

* Correspondence: bistra.stamboliyska@orgchm.bas.bg

Abstract: This presented research examined the wall paintings in the Church-Ossuary Presentation of the Blessed Virgin, part of the most important Rila Monastery complex in Bulgaria, painted by the painters from Mount Athos in 1795. The painting materials used to create the unique murals were studied for the first time by optical microscopy (OM), attenuated total reflectance Fourier transform infrared (ATR–FTIR), scanning electron microscopy energy dispersive X-ray spectroscopy (SEM–EDS), and enzyme-linked immunosorbent assay (ELISA). The vibrational and elemental analysis showed that the color palette of the paintings is composed of pigments traditional for Orthodox church wall paintings such as natural pigments, including yellow ochre, red ochre, green earth, and calcite, as well as other historical pigments of synthetic origin, including smalt, red lead, cinnabar, and verdigris. The analysis of the binders by the ATR–FTIR spectroscopy and enzyme-linked immunosorbent assay (ELISA) analysis implied the use of the post-Byzantine egg tempera technique. Only the blue backgrounds in the murals were painted using a smalt-based paint mixed with a carbohydrate binder. Based on the current analysis and comparison with the successive paintings in the other churches of the Rila Monastery, it could be concluded that the technology of the painting process followed by the Athonite artists during the work in the Church-Ossuary became a point of reference for many Bulgarian icon-painters later.

Keywords: Rila Monastery; the Church-Ossuary; pigments; binder; ATR–FTIR; SEM–EDS; ELISA

1. Introduction

The Rila Monastery is leading spiritual center in Bulgaria, situated at 1150 m on the highest mountain of the Balkan Peninsula. The monastery was established in the X century and was of great importance in the cultural life of the Bulgarians during the entire period of its existence [1]. Even though it is difficult to travel to, the monastery used to attract crowds of visitors when Bulgaria was under Ottoman rule (a time when, unlike today, there were no roads or transport) and especially during the Bulgarian Revival (in the second half of the eighteenth century and the first half of the nineteenth century). Today, as a cultural and religious site dedicated to the most venerated Bulgarian saint St. Ivan Rilski, the monastery is visited by thousands of visitors every year [2]. The Rila Monastery was recognized as a World Cultural Value in 1983 when it was included on the World Heritage List [3]. The monastery complex consists of the main church and farming buildings, surrounded by a fortified stone wall (Figure 1a), as well as some other churches and convents, located around the pilgrimage route [4]. A notable part of the monastery is the ensemble of the Church-Ossuary (the cemetery church) which also holds great religious and artistic importance. The building is situated to the south of the monastery walls, near the River Manastirska [5] (Figure 1a). The church, dedicated to the Presentation of the Blessed Virgin, was built in the Middle Ages on the foundations of the old ossuary (Figure 1b). The ground storing of the current building comprises a rectangular room housing the monks' ossuary. On the upper store, there is a one-nave church with one apside, an altar space with five niches in the walls, and a long rectangular nave (Figure 1c).

(a) (b) (c)

Figure 1. The Rila Monastery and the Church-Ossuary: (**a**) an air view of the Rila Monastery complex with the location of the Church-Ossuary marked by a red circle; (**b**) a general view of the Church-Ossuary building, east facade; (**c**) the ossuary and the church upstairs, west facade.

The church was decorated with wall paintings, first in the Rila Monastery complex in 1795 by the Athonite masters Zacharius and Veniamin of Galatista [6], who at this time (the end of the 18th century) resided and worked in the Rila monastery. The work of the Athonite painters influenced the style of the most famous Bulgarian iconographers, who, during the Bulgarian National Revival, painted many Orthodox churches and led to the introduction of new themes and models in ecclesiastical painting in Bulgaria. The murals of the Church-Ossuary of the Rila Monastery are of a high quality and rank among the best examples of church painting in Bulgaria from the eighteenth century (Figure 2).

The mural decoration in the church is constantly exposed to the harsh conditions of the mountain climate which inevitably affects their preservation state. This requires timely measures for the preservation and restoration of the endangered murals.

Figure 2. Selected scenes from the wall paintings of the Church-Ossuary of the Rila Monastery: (**a**) Churchyard Presentation of the Blessed Virgin; (**b**) The Protecting Veil of the Blessed Virgin; (**c**) Christ the Angel of the Great Council; (**d**) Saint Paphnutius and Saint Macarius of Egypt.

The identification and analysis of art materials used in mural creation is essential for the preparation and determination of the most appropriate strategy for murals' restoration and conservation and to optimize the conditions for storing them. Vibrational spectroscopy is a commonly used method for art materials analysis as it can give general information about both the mineral and organic materials, including mortars, pigments, and binders in the painting materials. Since many of the minerals commonly used as red, orange, and yellow pigments are heavy metal oxides, sulfides, and hydroxides—for instance, red lead, hematite (present in red ochres), goethite and limonite (present in red and yellow ochres), cinnabar, massicot, and others—they absorb IR radiation in the spectral interval below 500 cm^{-1}. Therefore, the far IR analysis is particularly useful for the identification of these pigments [7]. Over the past few decades, there have been numerous publications in which the capabilities of vibrational spectroscopy have been employed in the investigation of a variety of mural paintings [7–11]. Unfortunately, due to the complex nature of painting materials (present in the wall paintings), in many cases their identification based solely on the vibrational spectra is challenging. This is because the characteristic bands of different components in the multicomponent mixtures frequently overlap, making it difficult to distinguish between them. For this reason, a combination of the vibrational study with SEM–EDS analysis is applied as a complementary method for the identification of the pigments in painting layers [12–15]. The identification of organic materials (binders) within paint samples is a more difficult task due to the typical smaller quantities of the organic materials compared to the inorganic ones (pigments and mortar). Furthermore, the deterioration processes that occur over time may further reduce the quantities of these organic materials. For the reliable and complete analysis of the binder, the spectral methods are usually combined with chromatography-/mass spectrometry-based techniques (GC/MS, Py-GC/MS, HPLC) [16–21], as well as enzyme-linked immunosorbent assay (ELISA) [22–28]. The latter technique is a highly specific and sensitive method for identification of proteins even in trace amounts.

Such a combined approach encompassing vibrational spectroscopy and other analytical techniques—elemental analysis (SEM–EDS), X-ray diffraction (XRD), thermogravimetric analysis (TGA), and enzyme-linked immunosorbent assay (ELISA)—has previously been successfully employed in the study of different parts of the mural decoration in the Rila Monastery. These include the main altar [29], the nave [30], the exterior wall paintings [31], and the chapel of St Nikolay [32] of the main church of the Rila monastery and the Church of the Dormition of St John of Rila [33], which were created by the Bulgarian religious artists of the period. Complementary analytical methods have been widely applied for the study of murals decorated during the same and earlier periods in the Balkan region [34–48], Italy [49–53], and Turkey [54–56].

The present study aims to characterize the mural painting materials from the Church-Ossuary of the Rila Monastery with analytical methods and to extend the knowledge about the technical features of the ecclesiastical wall paintings created by the Athonite painters in Bulgaria. This will permit us to better outline the development of religious art in Bulgaria at the boundary between the 18th and 19th centuries and the role of the Athonite painters in it. The technological information is of great importance for the assessment of the murals' state and their further restoration.

2. Materials and Methods

A series of micro samples from the representative colors in the mural decoration in the Church-Ossuary (Figure 3) were collected by the restoration team. Approximately 1–3 mg samples of paint were scratched as fine powder from the partially damaged areas (to avoid the further damage of the wall paintings' state). The collection of the scratched powder was restricted as much as possible to the paint layer, without the ground layer.

Figure 3. Saint Nicephorus and Saint Orestes. A wall painting on the west side of the nave from where the studied samples S1–S7 were collected.

An optical microscopy (OM) examination was carried out with a Stemi 305 stereozoom microscope completed unit with built-in LED lighting. The microscope has a manual zoom of 5:1 (0.8–4.0×) and eyepiece set of 10×/23. The working distance is 110 mm, the viewing angle is 45°, and it uses a ZEISS Axiocam microscope camera.

The morphology and qualitative elemental analyses were studied using JEOL JSM 6390 (Tokyo, Japan). The images were researched with detectors for secondary and backscattered electrons. The accelerating voltage was 20 kV. The EDS had a 10 mm^2 detector with a 140 eV resolution and was produced in Oxford (INCA Oxford Instruments, Abingdon, UK). The samples were studied as powders. Prior to the analysis, the surface of the samples was thinly gold-coated. For each sample, the elemental composition was examined at several points on the sample surface (the full data are presented in the Supplementary Materials).

The ATR–FTIR spectra were recorded using a Bruker Invenio R FT spectrometer at a resolution of 2 cm^{-1}, with 100 scans in the region 4000–100 cm^{-1}. The samples were studied with an ATR technique by direct application and pressing on the diamond ATR crystal.

For the enzyme-linked immunosorbent assay (ELISA), the samples (approx. 1 mg) were soaked in 0.5 mL sterile carbonate–bicarbonate buffer (pH 9.6, 50 mM) for 24 h. Then, the sample solutions (100 μL) were loaded on a 96-well Nunc maxisorb immuno plate (ThermoScientific, Waltham, MA, USA) and were incubated overnight at 4 °C. Then, to avoid false positive results, the unspecific sites of the wells were blocked by adding 100 μL of blocking buffer (Ultrablock buffer, Bio-Rad, Hercules, CA, USA) and gently shaking them for 2 h at room temperature. In two subsequent steps, for the detection of egg white, a rabbit anti-chicken ovalbumin primary antibody (1:10,000) (Bio-Rad) and a goat anti-rabbit IgG labeled with an alkaline phosphatase secondary antibody (1:2000) (Bio-Rad) were added. For the detection of casein, a rabbit polyclonal antibody to casein (1:500) (Abcam, Cambridge, UK) and the same goat anti-rabbit IgG labeled with an alkaline phosphatase secondary antibody (1:1500) were used. The samples were incubated with the optimized concentrations of the antibodies for 1 h at room temperature. Washing with 3 × 300 μL phosphate saline buffer (pH 7.4) containing 0.05% Tween 20 preceded each step. Finally, 100 μL of p-nitrophenyl phosphate (substrate) was added to each well and the plate was shaken for 30 min in a dark place. The reaction was stopped by adding 50 μL of 0.1 N sodium hydroxide. The OD at 405 nm of the ELISA plate was read on a HiPo Microplate reader (Biosan, Warren, MI, USA). Control experiments without antigens were performed and were considered a negative control. The results are presented as the means of three independent measurements.

3. Results and Discussion

In order to study the color palette and the technological aspects of the wall paintings in the church, the samples collected from the Church-Ossuary of the Presentation of the Blessed Virgin were analyzed with OM, FTIR, SEM–EDS, and ELISA techniques. A brief description of the color, the sampling areas, and the data obtained from the sample analysis is given in Table 1. The color and size of the paint grains were studied by optical microscopy. Relevant microscopic images of the powder samples from the paint layer are presented in Figure S1 in the Supplementary Materials. During the collection of the painted layer samples, it was revealed that the blue paint of the background was applied on top of a black underlayer, while in all the other cases, the paint was on top of a white ground.

To identify similarities and differences between the wall painting materials and techniques used by the Athonite masters in Church-Ossuary and those used by the Bulgarian painters in the main church of the Rila monastery, the data were compared with previous findings. The identification of the materials and technical specifics can permit us to obtain a fuller understanding of the painting process and its evolution from the Athonian traditions to the Bulgarian painting schools of the time.

Table 1. Elemental composition identified components and characteristic ATR–FTIR and Raman frequencies of studied samples.

Sample	Elemental Composition *	Identified Components	IR Bands (cm^{-1})
S1—Blue from background	O, Si, Co, K, As, Ca	Smalt	1010 vs, 784 m, 430 s
		Binder (carbohydrate glue)	3315 w, 2920 w, 2852 w, 1648 m, 1416 w, 1152 sh, 1019 s, 607 w
S2—Green from background	C, O, Cu, Ca, Pb	Verdigris	3460 w, 3360 w, 1606 m, 1447 sh,1366 m, 1034 w, 693 w, 629 w, 502 w, 360 w
		Calcite	1406 vs, 872 s, 712 m
		Binder (egg)	2924 w, 2856 w, 1735 w, 1640 m
S3—Green from garments	O, C, Ca, Si, Fe, Al, Mg, K	Green earths	3603 w, 3557 w, 3534 w, 1073 m, 960 s, 798 w, 746 w, 671 m, 482 m, 450 m, 279 m
		Calcite	1440 m, 878 w, 712 w
		Binder (egg)	2956 w, 2926 w, 1734 w, 1640 w
S4-Yellow from garments	O, C, Ca, Fe, Si, Al, Mg, K	Yellow ochre	3150 w, 1025 ms, 1116 s, 1029 m, 912 w, 797 w, 467 w
		Calcite	1420 s, 872 m, 712 w
		Binder (egg)	2924 w, 2854 w, 1638 w
S5—Red from garments	O, C, Fe, Ca, Hg, Si, S, Al, K	Red ochre	3694 w, 3624 w, 1094 m, 1029 m, 1010 m, 913 w, 664 w, 603 m, 532 w, 447 w
		Calcite	1406 vs, 872 s, 712 m
		Cinnabar	342 w, 285 w, 280 sh
		Binder (egg)	2922 w, 2856 w, 1735 w, 1638 w
S6—Orange from frames	O, C, Ca, Pb	Red lead	681 w, 525 w, 428 w, 426, 363 w, 308 w, 275 w
		Calcite	1396 vs, 872 s, 711 m
		Binder (egg)	3245 w, 2920 w, 2853 w, 1736 w, 1634 m
S7—White from frames	O, C, Ca, Cu, Si	Calcite	1406 s, 872 s, 712 m, 288 s
		Binder (egg)	2925 w, 2853 w, 1738 w, 1642 m

* Elements are given in decreasing amount (content in weight % is provided in the Supplementary Materials).

3.1. *Pigments Identified in the Background Samples*

The wall paintings of the church were executed according to Byzantine and post-Byzantine techniques. In the initial stage, the secondary elements, namely, the backgrounds, were created. Similarly to the Byzantine monuments [34], the background of the scenes is separated into two zones. The upper section features a larger blue zone symbolizing the sky and decorated by gilded stars. In the lower section, there is a zone of a green shade. In the spirit of medieval tradition, the blue background was applied on a black underlayer. This hagiography technique of painting a blue color on top of a black ground layer was mastered and applied half a century later by the Bulgarian painters in painting the background of the main church of the Rila monastery [29,30]. The black pigment in the main church of the Rila monastery was identified as a carbon black of plant origin based on its Raman spectrum, where the bands at 1580 and 1330 cm^{-1} were assigned to amorphous carbon [57]. In the present investigation, the blue pigment applied on top of the black layer was identified as smalt–cobalt glass ($CoO.nSiO_2$) by the help of IR spectroscopy and SEM–EDS analysis (Table 1, Figure 4). In the ATR–FTIR spectrum of the blue sample, shown in Figure 4b, all the characteristic bands of smalt [58], such as the strong band at 1010 cm^{-1} (assigned to Si–O stretching), the medium-intensity band at 784 cm^{-1} (assigned to Si–O bending), and the weak-intensity bands at 693 and 617 cm^{-1} (assigned to Co-O) are present. These data, supported by the detection of silicon (Si) and cobalt (Co) by the EDS analysis, confirmed the use of smalt. The backscatter electron image (Figure 4a) revealed particles of irregular shape, but smooth morphology and particles of smalt appear as opaque pieces of glass. Detailed data on the elemental compositions found at the different points of the blue sample and grain size are provided in Table S1 in the Supplementary Materials. Smalt is an artificial potash silicate pigment, whose shade (color) depends on the relative contents of CoO and SiO_2 [59,60]. The pigment was used frequently in church wall paintings in the post-Byzantine period [13,34,61], despite some known stability issues related to the discoloration and yellowing of the pigment [61,62].

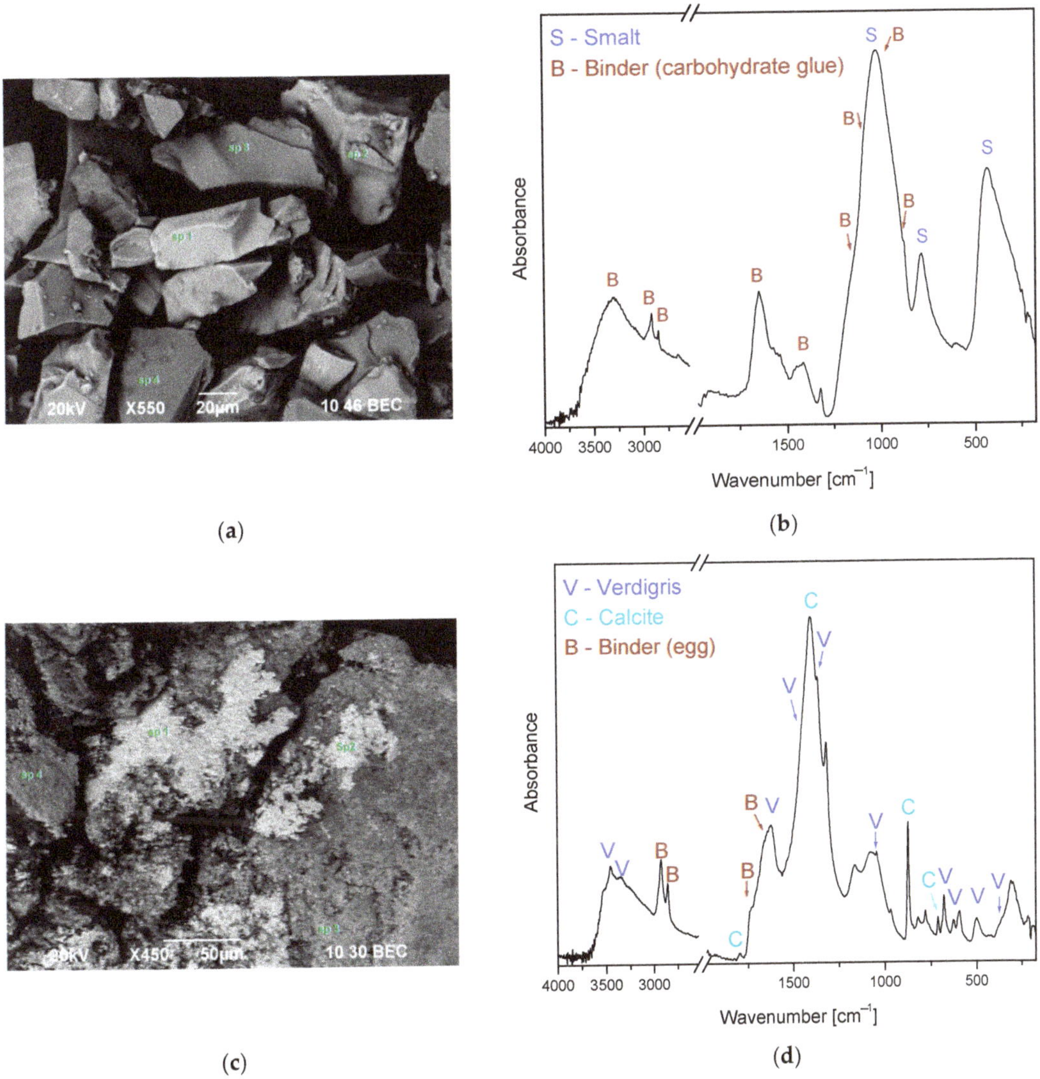

Figure 4. Backscattered electron images of the blue (**a**) and green (**c**) samples from the background; ATR–FTIR spectra of the blue (**b**) and green (**d**) samples from the background.

In the lower section of the background, the green paint was applied directly onto a white plaster. The SEM–EDS analysis of the sample revealed the presence of Cu in the elemental composition (Figure 4a, Table S2) and suggested the possible use of a copper pigment. Two copper pigments, malachite (basic copper(II) carbonate $CuCO_3.Cu(OH)_2$) and verdigris (basic copper acetate $(Cu(OH)_2(CH_3COO)_2.H_2O)$), were present in painters' palettes during the post-Byzantine period. They offered a more intense and saturated hue than the green earth mainly employed in the Byzantine times [34]. The appearance of the characteristic IR bands of the acetate group (COO-) in the middle IR region spectrum, at 1606 cm^{-1} (asymmetric C–O stretching), 1447 cm^{-1} (symmetric C–O stretching), and 690 cm^{-1} (C–O deformation band) [63], as well as characteristic bands in far the IR region 600–200 cm^{-1} [7] of the ATR–FTIR spectrum (Figure 4b, Table 1) of the green sample, clearly support the presence of verdigris. The verdigris pigment was employed frequently in wall paintings throughout the 19th century, as it was the most vibrant green pigment available at the time [63].

It is noteworthy that the green section in the background of the Church of the Dormition of St. John of Rila, painted between 1828 and 1830 [33], was created using verdigris, as seen in the Church-Ossuary. Meanwhile, in the backgrounds in the main church of the Rila monastery (painted later, between 1840 and 1842) [29–32], the painters used a new synthetic pigment, emerald (Schweinfurt) green, commercially available from 1814, to achieve a brighter color shade.

3.2. *Pigments Identified in the Garments*

The investigation of the pigments from the garments of the saints showed that they comprised mainly earth pigments, also called colored clays, natural mineral mixtures of silicate and clay minerals with different coloring agents.

The SEM–EDS analysis of the green sample from the garments revealed the elements Si, Al, Fe, Mg, K, and Ca (Figure 5a, Table S3). This result implied that the green pigment used to paint the background was a natural green earth pigment ($K[(Al,Fe^{3+})(Fe2^{+},Mg)],AlSi_3,Si_4)O_{10}(OH)_2$). In these natural pigments, the green chromophoric elements (Fe^{3+} and Fe^{2+}) are part of the structure of the clay minerals, celadonite, glauconite, etc. Their green color is determined by the ratio between the Fe^{3+} and Fe^{2+} ions in the mineral structure and by the presence of impurities [64]. Green earths have often been used as green pigments in Byzantine churches across the Balkan Peninsula [41–43]. The ATR–FTIR spectrum of the green sample from the garments (Figure 5b, Table 1) exhibited characteristic bands at 3530 and 3554 cm^{-1} (O–H stretching), 1100, 1070, and 971 cm^{-1} (Si–O stretching), and 800 and 681 cm^{-1} (R–O–H bending, where R represents the octahedral metal atoms (Al, K, Mg)), which were attributed to green earth and specifically to celadonite [65].

The investigation of the yellow and red samples from the garments revealed that they comprised natural ochres and also earth pigments consisting of aluminosilicate. However, in these cases, the coloring agents were non-clay, e.g., iron oxides and iron oxyhydroxides. Red and yellow earth pigments were often used in Byzantine and post-Byzantine hagiography because of their good coloring properties and their durability being unaffected by environmental factors, illumination, and chemical transformations [45,47,55,65,66]. In the present case, the yellow pigment was identified as yellow ochre, which is composed of silica and clay such as kaolinite. The pigment's color is derived from the iron oxyhydroxide mineral goethite (FeO(OH)) [66]. The performed SEM–EDS analysis (Figure 5c, Table S4) revealed that the sample contained Al and Si, the elements characteristic of silicates, and additionally Fe. In the ATR–FTIR spectrum of the yellow sample (Figure 5d, Table 1), the characteristic bands of kaolinite at 3600, 1116, 1029, and 912 cm^{-1} and the characteristic bands of goethite at 911 and 798 cm^{-1} were found [66].

The red pigment was identified as a mixture of red ochre and a bright red mercuric sulfide (cinnabar, HgS). The red ochre consists of clay and owes its color to the red mineral hematite (α-Fe_2O_3) [66]. The elemental analysis of the red samples detected Fe (suggesting the presence of hematite), Al and Si (indicating the presence of aluminosilicates), and Hg and S (indicating the presence of cinnabar) (Figure 5e, Table S5). In the ATR–FTIR spectrum of the red sample from the garments, the characteristic bands of kaolinite and hematite (at 532, 447, and 319 cm^{-1}) [66] and cinnabar (at 342, 285, and 280 cm^{-1}) [67] were present (Figure 5f, Table 1). The mixing of red ochre with mercuric sulfide (cinnabar) to produce a darker color is characteristic of post-Byzantine wall paintings [34].

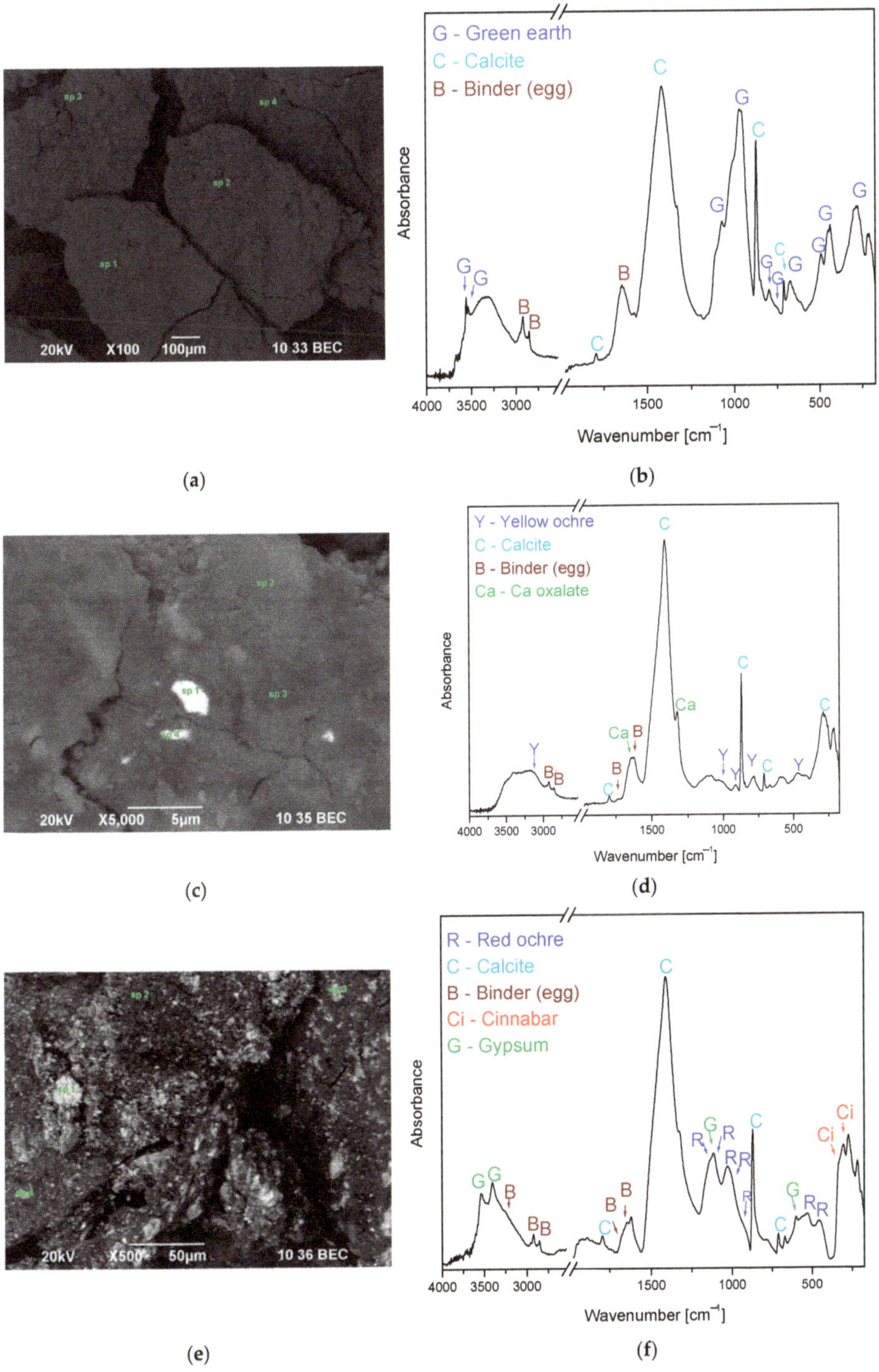

Figure 5. Back scattered electron images of the green (**a**), yellow (**c**), and red (**e**) samples from the garments; ATR–FTIR spectra of the green (**b**), yellow (**d**), and red (**f**) samples from the garments.

3.3. *Pigments Identified in the Frame*

Like the wall paintings in the main church of the Rila Monastery, the individual scenes and saints in the Church-Ossuary are surrounded by a frame of wide orange and thin white lines (Figures 2 and 3). To find out whether the Bulgarian painters used the same pigment as the Athonite masters, the orange and the white samples from the frames were studied. The SEM-EDX analysis of the orange sample (Figure 6a, Table S6) showed nothing but Pb in the elemental composition. The IR spectrum (Figure 6b, Table 1) showed clearly the characteristic peaks of minium (red lead, Pb_3O_4) [67,68]. Red lead is one of the first synthetically prepared pigments, used since ancient times, and despite its instability in humid environments, it was widely used in church painting due to its good hiding properties and low price [61].

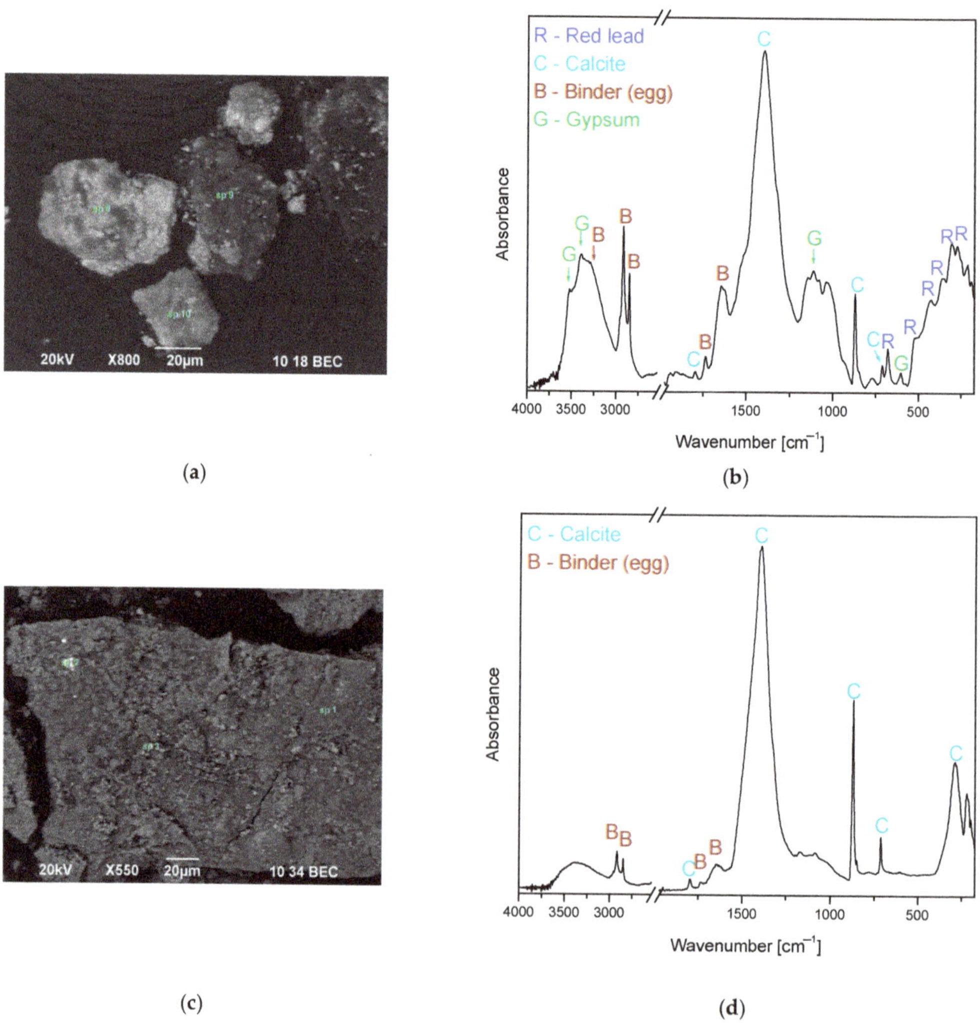

Figure 6. Back scattered electron images of the orange (**a**) and white (**c**) samples from the frame; ATR–FTIR spectra of the orange (**b**) and white (**d**) samples from the frame.

The white pigment from the frame, as confirmed by the SEM–EDS analysis (Figure 6c, Table S7), revealed mainly Ca, indicative of lime white (calcite). This result was supported by the ATR–FTIR spectrum of the sample shown in Figure 6c. In the spectra of the white paint samples collected from the frame are observed an intense broad band at 1425 cm^{-1} and a sharp band at 875 cm^{-1}, assigned to the asymmetric and symmetric stretching vibrations of the CO_3^{2-} ions, and a peak at 712 cm^{-1}, associated with the deformation vibration of the same group [69,70].

3.4. Painting Technique (Binder)

It is known that, in Byzantine and post-Byzantine ecclesiastic painting, several techniques for the application of binding media evolved. The first technique used lime as a binding medium (fresco technique), and an organic binding medium could also be applied (secco and mixed fresco–secco techniques). Technological studies of church wall paintings have established that organic binders based on proteins (including egg, animal glue, or casein), were the most commonly used. This is presumably because proteins, especially egg, have stronger bonding capacities than gums. According to the post-Byzantine technique, the painting commenced with the application of damp plaster and was subsequently completed using the secco technique with a protein binder [34,43].

Our ATR–FTIR analysis revealed that all of the examined samples exhibited a substantial concentration of calcium carbonate. However, IR bands in the 3000–2800 cm^{-1} region (associated with the C–H stretching vibrations of hydrocarbon chains) were also observed in all the spectra, indicating that an organic binder was also added to the paint. This absorption region is specific to organic compounds and does not overlap with those of inorganic materials, allowing the detection of the former even when they are present in smaller quantities. In addition to the bands for the C–H stretching vibrations, the careful examination of the ATR–FTIR spectra of the samples also revealed IR bands for the N–H stretching vibrations of amide groups (Amide A) at ca. 3300 cm^{-1}, for C–O stretching (Amide I) at 1630–1660 cm^{-1}, and N–H bending (Amide II) at 1500–1550 cm^{-1}, shoulder to the intense band of calcite (Figure 7). These IR bands correspond to protein material. The use of egg (whole egg or egg yolk) as a binder usually results in the appearance of a characteristic lipid band in the IR spectrum, situated at approximately 1730 cm^{-1} (C–O ester stretching) [71]. Moreover, the relative intensities of the lipid carbonyl stretching band and the amide I might indicate whether whole egg or only yolk was used based on the different relative amounts of the lipids in both cases [53,72]. The intensity ratio of the two bands in dried whole egg, yolk, and egg white (lipid band is missing) is illustrated in Figure S2 in the Supplementary Materials.

Our IR analysis showed that two of the paint samples (green and red from the garments) lacked a distinct lipid carbonyl stretching band, but in the other cases, the ATR–FTIR spectra contained a band at such a position, although it was of a weak intensity. The ATR–FTIR spectra of the samples are compared with the spectrum of the reference aged, dried egg in Figure 7. Where the lipid band is observed, the intensity ratio of the amide I and the lipid carbonyl stretching band corresponds to those characteristic for the whole egg (Figure S3 in the Supplementary Materials), which indicates that, most probably, the binder is whole egg.

To ascertain the use of egg as a binder, the analysis was complemented by enzyme-linked immunosorbent assay (ELISA). The samples were tested with the antibodies of egg white and casein. The results obtained are shown in Figure 8, where the dashed line represents the limit value of the analysis (the lowest limit of detection). The results clearly confirm the presence of egg white in the paint, as the optical density of all the tested samples was above the cut-off of the assay for egg white (Figure 8a). None of the tested samples did not show an optical density above the cut-off value in the casein assay (Figure 8b).

As a specific test for animal glue was also not conducted, its presence cannot not be completely ruled out. However, due to the established intensity ratio corresponding to whole egg, it might be concluded that egg is most probably the predominant part. A mixed protein binder containing predominantly whole egg and a smaller portion (below 2%) of animal glue was discovered in the study of the Protaton Church on Mount Athos, Greece [43].

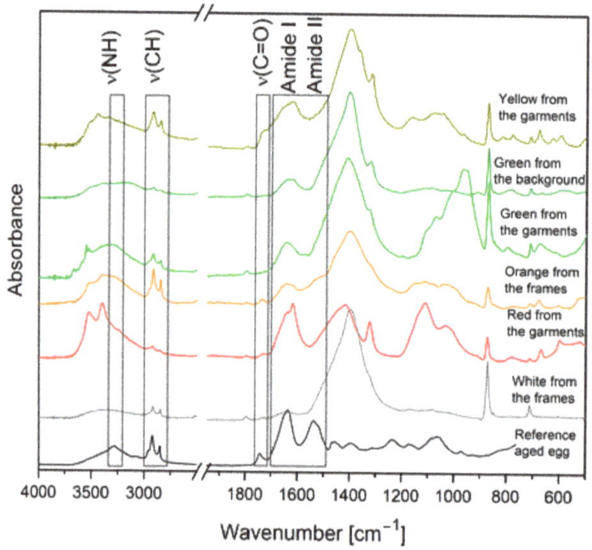

Figure 7. ATR–FTIR spectra of the studied samples compared with the reference aged egg.

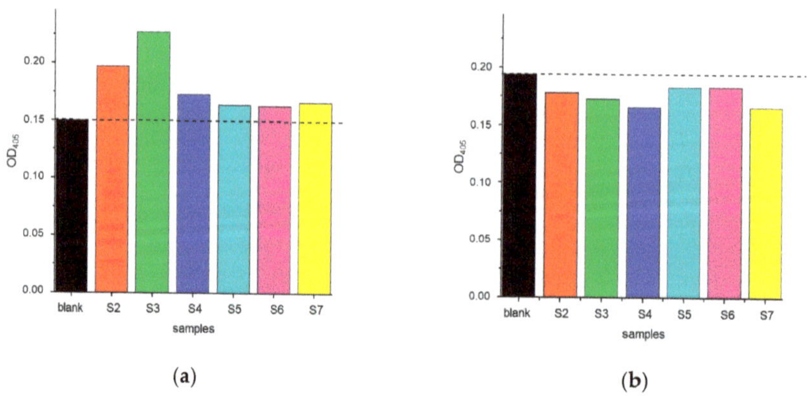

Figure 8. Detection of egg white (**a**) and casein (**b**) by the measurement of the optical density at 405 nm using ELISA at optimized antibody dilutions. Conditions for egg white detection: rabbit anti-chicken ovalbumin (primary antibody, 1:10,000) and goat anti-rabbit IgG alkaline phosphatase (secondary antibody, 1:2000). Conditions for casein detection: rabbit polyclonal antibody to casein (primary antibody, 1:500) and goat anti-rabbit IgG alkaline phosphatase (secondary antibody, 1:1500). The dashed line represents the assay cut-off value. For sample numbering, see Table 1.

The substantial amount of calcium carbonate in the painted layer is evidence that the pigment was applied in a water–lime solution on wet plaster. During the drying of the plaster and paint layer, carbonation takes place and, as a result, the pigment is tightly bound to the plaster. On the other hand, an organic binder is also present, which indicates that a mixed fresco–secco mural technique was used.

Unlike the other samples, the ATR–FTIR spectrum of the blue sample did not show the presence of protein. When comparing the spectrum with the spectra of various organic binders, the best agreement was obtained with the spectrum of wheat carbohydrate glue. The following IR bands, related to carbohydrates, were observed: 3310 cm^{-1} (O–H stretching vibrations); 2920 and 2852 cm^{-1} (C–H stretching vibrations); 1648 cm^{-1} (O–H bending vibrations); 1416 cm^{-1} (C–H bending vibrations); and 1019 cm^{-1} (C–O stretching vibrations). A comparison of the ATR–FTIR spectrum of the blue sample and reference carbohydrate glue is presented in Figure S4 in the Supplementary Materials. No IR bands for calcium carbonate were observed in the spectrum, suggesting that, contrary to the other pictorial layers, the blue paint layer was applied on dry plaster.

The use of smalt pigment in egg binding media has been reported for wall paintings in Greek and Cypriot monasteries [34,35,61]. On the other hand, it was also documented that smalt might interact with egg binder, and its color could change to green–grey hues as a result [73]. This is probably the reason why the blue background zones in the wall paintings in the Catholicon of the Rila Monastery, which were much later created, showed exclusively carbohydrate binders. It is noteworthy that, in the wall paintings of the Church-Ossuary, preceding the Catholicon of the Rila Monastery, the blue background was also painted by the Athonite painters with a carbohydrate binder.

4. Conclusions

The ATR–FTIR, SEM–EDS, ELISA study of the mural paintings of the Church-Ossuary Presentation of the Blessed Virgin provided significant insights into the materials and techniques used by the painters from Mount Athos in their work in the Rila Monastery, one of the most important spiritual centers in Bulgaria. Their ecclesiastical wall paintings influenced the works of the Bulgarian icon painters which later created the polychromy in the main church in the Rila monastery.

Through the application of instrumental analytical investigations, the following similarities in the work of the Athonite painters and their later Bulgarian followers were found:

- The mixed fresco–secco mural technique was used, which did not rely solely on the binding properties of the carbonizing lime, but the paint was also mixed with a protein binder (most probably egg) during the working process.
- The colors in the figures and garments were based on earth pigments—green earths, yellow ochre, and red ochre—which have a long history of being used in Orthodox church wall painting.
- The blue pigment used for representing the sky was smalt, applied in a dense paste-like layer on top of an earlier prepared black underlayer at the final stage of the painting, in the spirit of medieval tradition. The smalt pigment was mixed with a carbohydrate binder, unlike other Greek and Cypriot church wall paintings where egg has been found.

On the other hand, some aspects of the painting technique and aesthetic features differ in the work of the Athonite and Bulgarian painters:

- The background in the scenes in the Church-Ossuary is separated in two zones (blue and green), while there are four zones in the Rila monastery catholicon with a different color scheme (blue, pink, yellow, and green).
- In both cases, the pigments used for painting the background colors are of synthetic origin and provide more vivid colors than those of the garments. However, while in the Church-Ossuary the green paint is based on verdigris, which was known for a long time before the creation of the wall paintings, it is replaced in the Rila monastery Catholicon by a modern green pigment (emerald green). Another modern pigment, chrome yellow, is also introduced as a yellow color in the backgrounds of the Catholicon.

The current study improves the understanding of the technological diversity of post-Byzantine wall paintings and their evolution in Bulgarian churches. It could be concluded that many of the technological features—the binding media and pigments—of the Athonite masters were transferred to the work of the Bulgarian icon painters, and at the same time, some of them evolved into more modern materials.

Supplementary Materials: The following supporting information can be downloaded at: https://www.mdpi.com/article/10.3390/min14111115/s1, Table S1: EDS results of the blue sample from the background; Table S2: EDS results of the green sample from the background; Table S3: EDS results of the green sample from the garments; Table S4: EDS results of the yellow sample from the garments; Table S5: EDS results of the red sample from the garments; Table S6: EDS results of the orange sample from the frame; Table S7: EDS results of the white sample from the frame; Figure S1: OM images of the studied sample with the measured size of paint grains; Figure S2: FTIR spectra of dried whole egg, egg white, and egg yolk; Figure S3: FTIR spectra of dried whole egg, dried yolk egg, white paint from the frame, and red from the frame; Figure S4: FTIR spectra of the blue sample from the background and carbohydrate glue.

Author Contributions: Conceptualization, B.S., S.T., M.G., B.R. and D.Y.; Formal analysis, E.V., S.A.-V., M.G., S.S. and D.Y.; Funding acquisition B.S. and B.R.; Investigation, B.S., E.V., S.A.-V., B.R., M.G., S.S. and D.Y.; Methodology, B.S., S.T., S.A.-V., B.R., M.G. and D.Y.; Validation, B.S., E.V., S.A.-V., B.R., M.G., S.S. and D.Y.; Visualization, E.V., S.A.-V., S.S., and M.G.; Writing—original draft, B.S., S.A.-V., M.G. and D.Y.; Writing—review and editing, B.S., E.V., S.A.-V., B.R., M.G., S.S. and D.Y. All authors have read and agreed to the published version of the manuscript.

Funding: This work has been financially supported by Project INFRAMAT, part of the Bulgarian National Roadmap for Scientific Infrastructure, supported by the Ministry of Education and Science. D.Y. is grateful for the financial support of the European Union NextGenerationEU, through the National Recovery and Resilience Plan of the Republic of Bulgaria, Project No. BG-RRP-2.004-0002, "BiOrgaMCT". B.R. and S.A.V. acknowledge the support by the European Regional Development Fund within the OP Science and Education for Smart Growth 2014–2020 and Project CoE National centre for mechatronics and clean technologies, No. BG05M2OP001-1.001-0008.

Data Availability Statement: The original data presented in the study are included in the article, further inquiries can be directed to the corresponding authors.

Conflicts of Interest: The authors declare no conflicts of interest.

References

1. Terziyska, M. Cultural and educational role of the Rila Monastery in the history of Bulgaria (X–XIX centuries). A bridge between eras, generations and cultures. *Hist. Educ. Child. Lit.* **2016**, *11*, 11–25.
2. Tonev, L. The Rila Monastery Historical monument and national museum. *Museum Int.* **1978**, *30*, 109–115. [CrossRef]
3. UNESCO. Report on the 7th Session of the Communities. 1983. Available online: http://whc.unesco.org/en/sessions/07COM/ (accessed on 20 March 2020).
4. Kuyumdzhiev, A.; Mutafov, E.; Vanen, I.; Gergova, I.; Popova, E.; Genova, E.; Uyumdzhieva, M.; Zaharieva, M.; Stoykova, M.; Christemova, M.; et al. *Corpus of Mural Paintings from the First Half of 19th Century in Bulgaria*; Institute of Art Studies, Bulgarian Academy of Sciences: Sofia, Bulgaria, 2018.
5. Koeva, M. *A Cultural Value Consigned to Forgotten—the Cemetery Church of the Rila Monastery (Original Title: Edna Kulturna Cennost, Predadena na Zabvenie-Ansambalat na Grobitschnsta Cherkva na Rilskia Manastir)*; LiterNet: Varna, Bulgaria, 2003. Available online: https://liternet.bg/publish9/mkoeva/nasledstvo/grobishtna_cyrkva.htm (accessed on 20 September 2024).
6. Kuyumdzhiev, A. Mural Paintings of Athonite Icon Painters in Rila Monastrey from the end of the 18th Century. In *Art Readings 2019*; Moutafov, E., Kuyumdzhieva, M., Eds.; Institute of Art Studies, Bulgarian Academy of Sciences: Sofia, Bulgaria, 2020; pp. 413–438.
7. Vahur, S.; Teearu, A.; Leito, I. ATR-FT-IR spectroscopy in the region of 550–230cm^{-1} for identification of inorganic pigments. *Spectrochim. Acta Part A Mol. Biomol. Spectrosc.* **2010**, *75*, 1061–1072. [CrossRef] [PubMed]
8. Bitossi, G.; Giorgi, R.; Mauro, M.; Salvadori, B.; Dei, L. Spectroscopic Techniques in Cultural Heritage Conservation: A Survey. *Appl. Spectrosc. Rev.* **2005**, *40*, 187–228. [CrossRef]
9. Doménech Carbó, M.T.; Bosch Reig, F.; Gimeno Adelantado, J.V.; Periz Martínez, V. Fourier transform infrared spectroscopy and the analytical study of works of art for purposes of diagnosis and conservation. *Anal. Chim. Acta* **1996**, *330*, 207–215. [CrossRef]

10. Rosi, F.; Daveri, A.; Miliani, C.; Verri, G.; Benedetti, P.; Pique, F. Non-invasive identification of organic meterials in wall paintings by fiber optic reflectance infrared spectroscopy: A statistical multivariate approach. *Anal. Bioanal. Chem.* **2009**, *95*, 2097–2106. [CrossRef]
11. Mangone, A.; Colombi, C.; Eramo, G.; Muntoni, I.M.; Forleo, T.; Giannossa, L.C. Pigments and Techniques of Hellenistic Apulian Tomb Painting. *Molecules* **2023**, *28*, 1055. [CrossRef]
12. Iordanidis, A.; Garcia-Guinea, J.; Strati, A.; Gkimourtzina, A.; Papoulidou, A. Byzantine wall paintings from Kastoria, northern Greece: Spectroscopic study of pigments and efflorescing salts. *Spectrochim. Acta Part A Mol. Biomol. Spectrosc.* **2011**, *78*, 874–887. [CrossRef]
13. Danillia, S.; Minopoulou, E.; Demosthenous, D.; Karagiannis, G. A comparative study of wall paintings at the Cypriot monastery of Christ Antiphonitis: One artist or two? *J. Archaeol. Sci.* **2008**, *35*, 1695–1707. [CrossRef]
14. Zorba, T.; Andrikopoulos, K.S.; Paraskevopoulos, K.M.; Pavlidou, E.; Popkonstantinov, K.; Kostova, R.; Platnyov, V.; Daniilia, S. Infrared and Raman Vibrational Spectroscopies Reveal the Palette of Frescos Found in the Medieval Monastery of Karaach Teke. *Ann. Chim.* **2007**, *797*, 491–503. [CrossRef]
15. Sakellariou, E.; Zorba, T.; Pavlidou, E.; Angelova, S.; Paraskevopoulos, K.M. The byzantine church of "40 holy martyrs" in Veliko Turnovo, Bulgaria: Pigments and technique. *AIP Conf. Proceed.* **2010**, *1203*, 501–506.
16. Mazurek, J.; Svoboda, M.; Schilling, M. GC/MS Characterization of Beeswax, Protein, Gum, Resin, and Oil in Romano-Egyptian Paintings. *Heritage* **2019**, *2*, 1960–1985. [CrossRef]
17. Cirlini, M.; Graiff, C.; Lottici, P.P.; Palla, G.; Casoli, A.; Bergamonti, L. Characterization of Waxes in the Roman Wall Paintings of the Herculaneum Site (Italy). *Appl. Sci.* **2022**, *12*, 11264. [CrossRef]
18. Brecoulaki, H.; Zaitoun, C.; Stocker, S.R.; Davis, J.L.; Karydas, G.; Colombini, M.P.; Bartolucci, U. An Archer from the Palace of Nestor: A New Wall-Painting Fragment in the Chora Museum. *Hesperia* **2008**, *77*, 363–397. [CrossRef]
19. Zumbühl, S.; Caruso, O.; Gammaldi, N.; Iazurlo, P.; Piqué, F. The Characterization of the Materials Used by Gino Severini in his 20th C Wall Paintings at Semsales in Switzerland Patrizia Moretti. *Appl. Sci.* **2021**, *11*, 9161.
20. da Filicaia, E.G.; Evershed, R.P.; Peggie, D.A. Review of recent advances on the use of mass spectrometry techniques for the study of organic materials in painted artworks. *Anal. Chim. Acta* **2023**, *1246*, 340575. [CrossRef]
21. Bonaduce, I.; Andreotti, A. Py-GC/MS of Organic Paint Binders. In *Organic Mass Spectrometry in Art and Archaeology*; Colombini, M.P., Modugno, F., Eds.; Wiley: Chichester, UK, 2009; Chapter 11; pp. 303–326.
22. Potenza, M.; Sabatino, G.; Giambi, F.; Rosi, L.; Papini, A.M.; Dei, L. Analysis of egg-based model wall paintings by use of an innovative combined dot-ELISA and UPLC-based approach. *Anal. Bioanal. Chem.* **2013**, *405*, 691–701. [CrossRef]
23. Li, J.; Zhanga, B. Study of identification results of proteinous binding agents in Chinese painted cultural relics. *J. Cult. Herit.* **2020**, *43*, 73–79. [CrossRef]
24. Cartechini, L.; Vagnini, M.; Palmieri, M.; Pitzurra, L.; Mello, T.; Mazurek, J.; Chiari, G. Immunodetection of Proteins in Ancient Paint Media. *Acc. Chem. Res.* **2010**, *43*, 867–876. [CrossRef]
25. Sun, X.; Zhang, B.; Chen, W. High-performance enrichment of casein and ovalbumin in wall painting with immunomagnetic beads. *Archaeometry* **2023**, *65*, 1322–1335. [CrossRef]
26. Zhang, B.; Chen, W. Electrochemical Chip Combined with Immunomagnetic Beads Enrichment for the Detection of Peach Gum Binding Medium in Ancient Wall Paintings. *Anal. Lett.* **2023**, *56*, 2869–2881.
27. Palmieri, M.; Vagnini, M.; Pitzurra, L.; Brunetti, B.G.; Cartechini, L. Identification of animal glue and hen-egg yolk in paintings by use of enzyme-linked immunosorbent assay (ELISA). *Anal. Bioanal. Chem.* **2013**, *405*, 6365–6637. [CrossRef]
28. Freeman, A.A.; de Ferri, L.; Mazurek, J.; Andriulo, F.; Bertolin, C. A Multi-Analytical Approach for the Characterization of Seventeenth Century DecorativeWall Paintings in Two Norwegian Stave Churches: A Case Study at Eidsborg and Heddal, Norway. *Appl. Sci.* **2021**, *11*, 3477. [CrossRef]
29. Stamboliyska, B.; Tapanov, S.; Velcheva, E.; Yancheva, D.; Rogozherov, M.; Glavcheva, Z.; Lalev, G.; Dimitrov, M. The altar wall paintings of the catholicon "The Nativity of the Virgin", Rila Monastery, Bulgaria: Identification of the painting materials bymeans of vibrational spectroscopic techniques complemented by EDX, XRD and TGA analysis. *Spectrochim. Acta Part A Mol. Biomol. Spectrosc.* **2021**, *247*, 119087. [CrossRef]
30. Yancheva, D.; Tapanov, S.; Velcheva, E.; Stamboliyska, B.; Glavcheva, Z.; Stoyanov, S.; Haralampiev, N.; Fischer, D.; Lederer, A. Characterization of Zahari Zograph's nave wall paintings in the church "Nativity of the Virgin" of Rila Monastery (Bulgaria) by vibrational spectroscopy and SEM–EDX analysis. *Sci. Techn. Arch. Res.* **2017**, *3*, 437–449. [CrossRef]
31. Stamboliyska, B.; Tapanov, S.; Kovacheva, D.; Atanasova-Vladimirova, S.; Ranguelov, B.; Yancheva, D.; Velcheva, E.; Stoyanov, S.; Guncheva, M.; Fischer, D.; et al. Characterization of art materials and degradation processes in the exterior wall paintings of the main church of Rila Monastery, Bulgaria. *Vibr. Spectr.* **2023**, *128*, 103580. [CrossRef]
32. Yancheva, D.; Tapanov, S.; Anastasova, N.; Velcheva, E.; Stoyanov, S.; Stamboliyska, B. Pigments and organic binders in the wall paintings of chapel "St. Nikolay" in Rila Monastery cathalicon, Bulgaria. *C. R. Acad. Bulg. Sci.* **2019**, *72*, 1468–1474.
33. Stamboliyska, B.; Tapanov, S.; Velcheva, E.; Yancheva, D.; Stoyanov, S. Spectroscopic Analysis of Wall Paintings from the Church of the Dormition of St John of Rila. *Rila Monast. J. Bul. Acad. Sci.* **2021**, *4*, 29–36.
34. Danillia, S.; Minopoulou, E.; Andrikopoulos, K.S.; Tsakalof, A.; Bairachtari, K. From Byzantine to post-Byzantine art: The painting technique of St Stephen's wall paintings at Meteora. *J. Am. Sci.* **2008**, *35*, 2474–2485. [CrossRef]

35. Daniilia, S.; Sotiropoulou, S.; Bikiaris, D.; Salpistis, C.; Karagiannis, G.; Chryssoulakis, Y.; Priced, B.A.; Carlson, J.H. Panselinos' Byzantine wall paintings in the Protaton Church, Mount Athos, Greece: A technical examination. *J. Cult. Herit.* **2000**, *1*, 91–110. [CrossRef]
36. Hein, A.; Karatasios, I.; Mourelatos, D. Byzantine wall paintings from Mani (Greece): Microanalytical investigation of pigments and plasters. *Anal. Bioanal. Chem.* **2009**, *395*, 2061–2071. [CrossRef] [PubMed]
37. Malletzidou, L.; Zorba, T.T.; Patsiaoura, D.; Lampakis, D.; Touli, V.; Chrissafis, K.; Karapanagiotis, I.; Pavlidou, E. Unraveling the materials and techniques of post-Byzantine wall paintings: Is there a sole pictorial phase at the catholicon of Stomion, Central Greece? *Spectrochim. Acta Part A Mol. Biomol. Spectrosc.* **2019**, *206*, 328–339. [CrossRef] [PubMed]
38. Mohanu, I.; Mohanu, D.; Gomoiu, I.; Barbu, O.-H.; Fechet, R.-M.; Vlad, N.; Voicu, G.; Truşcă, R. Study of the frescoes in Ioneştii Govorii wooden church (Romania) using multi-technique investigations. *Microchem. J.* **2016**, *126*, 332–340. [CrossRef]
39. Pavlidou, E.; Arapi, M.; Zorba, T.; Anastasiou, M.; Civici, N.; Stamati, F.; Paraskevopoulos, K. Onoufrios, the famous XVI's century iconographer, creator of the "Berati School": Studying the technique and materials used in wall paintings of inscribed churches. *Appl. Phys.* **2006**, *83*, 709–717. [CrossRef]
40. Holclajtner-Antunović, I.; Stojanović-Marić, M.; Bajuk-Bogdanović, D.; Žikić, R.; Uskoković-Marković, S. Multi-analytical study of techniques and palettes of wall paintings of the monastery of Žiča, Serbia. *Spectrochim. Acta Part A Mol. Biomol. Spectrosc.* **2016**, *156*, 78–88. [CrossRef]
41. Cheilakou, E.; Troullinos, M.; Koui, M. Identification of pigments on Byzantine wall paintings from Crete (14th century AD) using non-invasive Fiber Optics Diffuse Reflectance Spectroscopy (FORS). *J. Archaeol. Sci.* **2014**, *41*, 541–555. [CrossRef]
42. Danillia, S.; Tsakalof, A.; Bairachtari, K.; Chryssoulakis, Y. The Byzantine wall paintings from the Protaton Church on Mount Athos, Greece: Tradition and science. *J. Archaeol. Sci.* **2007**, *34*, 1971–1984. [CrossRef]
43. Zorba, T.; Pavlidou, E.; Stanojlović, M.; Bikiaris, D.; Paraskevopoulos, K.M.; Nikolić, V.; Nikolić, P.M. Technique and palette of XIIIth century painting in the monastery of Mileševa. *Appl. Phys. A Mater. Sci. Process.* **2006**, *8*, 719–725. [CrossRef]
44. Damjanović, L.; Marjanović, O.; Marić-Stojanović, M.; Andrić, V.; Mioč, U.B. Spectroscopic investigation of two Serbian icons painted on canvas. *J. Serb. Chem. Soc.* **2015**, *80*, 805–817. [CrossRef]
45. Marić-Stojanović, M.; Bajuk-Bogdanović, D.; Uskoković-Marković, S.; Holclajtner-Antunović, I. Spectroscopic analysis of XIV century wall paintings from Patriarchate of Peć Monastery, Serbia. *Spectrochim. Acta Part A Mol. Biomol. Spectrosc.* **2018**, *191*, 469–477. [CrossRef]
46. Malletzidou, L.; Zorba, T.; Kyranoudi, M.; Mastora, P.; Karfaridis, D.; Vourlias, G.; Pavlidou, E.; Paraskevopoulos, K. The dome of Rotunda in Thessaloniki: Investigation of a multi-pictorial phase wall painting through analytical methods. *Spectrochim. Acta Part A Mol. Biomol. Spectrosc.* **2021**, *262*, 120101. [CrossRef] [PubMed]
47. Marutoiu, C.; Bratu, I.; Trifa, A.; Botis, M.; Marutoiu, V.C. FTIR analysis of painting materials from the church Saint Paraschiva, of Poienile Izei, Maramureş, Romania. *Int. J. Conserv. Sci.* **2011**, *2*, 29–39.
48. Civici, N.; Anastasiou, M.; Zorba, T.; Paraskevopoulos, K.M.; Dilo, T.; Stamati, F.; Arapi, M. Studying wall paintings in Berati Castle (Albania): Comparative examination of materials and techniques in XIVth and XVIth century churches. *J. Cult. Herit.* **2008**, *9*, 207–213. [CrossRef]
49. Vasco, G.; Serra, A.; Manno, D.; Buccolieri, G.; Calcagnile, L.; Buccolieri, A. Investigations of byzantine wall paintings in the abbey of Santa Maria di Cerrate (Italy) in view of their restoration. *Spectrochim. Acta Part A* **2020**, *239*, 118557. [CrossRef] [PubMed]
50. Alberghina, M.F.; Zicarelli, M.A.; Randazzo, L.; Rigaglia, D.; Ricca, M. Byzantine wall paintings from San Marco d'Alunzio, Sicily: Non-invasive diagnostics and microanalytical investigation of pigments and plasters. *Heritage Sci.* **2024**, *12*, 184. [CrossRef]
51. Ospitali, F.; Rattazzi, A.; Colombini, M.P.; Andreotti, A.; Lonardo, G. XVI century wall paintings in the "Messer Filippo" cell of the tower of Spilamberto: Microanalyses and monitoring. *J. Cult. Herit.* **2007**, *8*, 323–327. [CrossRef]
52. Guglielmi, V.; Andreoli, M.; Comite, V.; Baroni, A.; Fermo, P. The combined use of SEM-EDX, Raman, ATR-FTIR and visible reflectance techniques for the characterisation of Roman wall painting pigments from Monte d'Oro area (Rome): An insight into red, yellow and pink shades. *Envir. Sci. Poll. Res.* **2022**, *29*, 29419–29437. [CrossRef]
53. Gelzo, M.; Grimaldi, M.; Vergara, A. Comparison of binder compositions in Pompeian wall painting styles from Insula Occidentalis. *Chem. Cen. J.* **2024**, *8*, 65. [CrossRef]
54. Bakiler, M.; Kırmızı, B.; Öztürk, Ö.O.; Hanyalı, Ö.B.; Çağlar, E.; Köroğlu, G. Material characterization of the Late Roman wall painting samples from Sinop Balatlar Church Complex in the black sea region of Turkey. *Microchem. J.* **2016**, *126*, 263–273. [CrossRef]
55. Demir, S.; Şerifaki, K.; Böke, H. Execution technique and pigment characteristics of Byzantine wall paintings of Anaia Church in Western Anatolia. *J. Archaeol. Sci. Rep.* **2018**, *17*, 39–46. [CrossRef]
56. Pelosi, C.; Agresti, G.; Andaloro, M.; Baraldi, P.; Pogliani, U.; Santamaria, M.F.; La Russa, S.A.; Ruffolo, N. Rovella, Micro-Raman and micro-stratigraphic analysis of P. the painting materials in the rock-hewn church of the Forty Martyrs in Şahinefendi, Cappadocia (Turkey). *Archaeometry* **2016**, *58*, 659–672. [CrossRef]
57. Burgio, L.; Clark, R.J.H. Library of FT-Raman spectra of pigments, minerals, pigment media and varnishes, and supplement to existing library of Raman spectra of pigments with visible excitation. *Spectrochim. Acta Part A Mol. Biomol. Spectrosc.* **2001**, *57*, 1491–1521. [CrossRef] [PubMed]
58. Jonynaite, D.; Senvaitiene, J.; Beganskiene, A.; Kareiva, A. Spectroscopic analysis of blue cobalt smalt pigment. *Vibr. Spectrosc.* **2010**, *52*, 158–162. [CrossRef]

59. Bouquillon, A.; Katona, I.; D'Allesandro, A.; Zucchiatti, A. PIXE analysis of artistic blue glazes and the use of cobalt pigments in ceramics during the Italian Renaissance. *Int. J. PIXE* **2005**, *15*, 317–322. [CrossRef]
60. Rehren, T. Aspects of the production of cobalt-blue glass in Egypt. *Archaeometry* **2001**, *43*, 483–489. [CrossRef]
61. Daniilia, S.; Minopoulou, E. A study of smalt and red lead discolourcoloration in Antiphonitis wall paintings in Cyprus. *Appl. Phys. A Mater. Sci. Process.* **2009**, *96*, 701–711.
62. Muhlethaler, B.; Thissen, J. Smalt. In *Artists' Pigments: A Handbook of Their History and Characteristics*; Roy, A., Ed.; Washington Archetype Publications: London, UK, 1993; Volume 2, pp. 113–130.
63. Kuhn, H. Verdigris and Copper Resinate. In *Artists' Pigments: A Handbook of their History and Characteristics*; Roy, A., Ed.; Washington Archetype Publications: London, UK, 1993; Volume 2, pp. 131–158.
64. Hradil, D.; Grygar, T.; Hradilova, J.; Bezdicka, P. Clay and iron oxide pigments in the history of painting. *App. Clay Sci.* **2003**, *22*, 223–236. [CrossRef]
65. Ospitali, F.; Bersani, D.; Di Lonardo, G.; Lottici, P.P. Green earths': Vibrational and elemental characterization of glauconites, celadonites and historical pigments. *J. Raman Spectrosc.* **2008**, *39*, 1066–1073. [CrossRef]
66. Bikiaris, D.; Daniilia, S.; Sotiropoulou, S.; Katsimbiri, O.; Pavlidou, E. Ochre-differentiation through micro-Raman and micro-FTIR spectroscopies: Application on wall paintings at Meteora and Mount Athos, Greece. *Spectrochim. Acta Part A Mol. Biomol. Spectrosc.* **1999**, *56*, 3–18. [CrossRef]
67. Vahura, S.; Knuutinenb, U.; Leito, I. ATR-FT-IR spectroscopy in the region of 500–230cm^{-1} for identification of inorganic red pigments. *Spectrochim. Acta Part A Mol. Biomol. Spectrosc.* **2009**, *73*, 764–771. [CrossRef]
68. Vigouroux, J.P.; Husson, E.; Calvarin, G.; Dao, N.Q. Etude par spectroscopic vibrationnelle des oxydes Pb_3O_4, $SnPb_2O_4$ et $SuPb(Pb_2O_4)_2$. *Spectrochim. Acta Part A Mol. Biomol. Spectrosc.* **1982**, *38*, 393–398. [CrossRef]
69. Gunasekaran, S.; Anbalagan, G. Spectroscopic characterization of natural Calcite Minerals. *Spectrochim. Acta Part A Mol. Biomol. Spectrosc.* **2007**, *68*, 656–664. [CrossRef] [PubMed]
70. Gunasekaran, S.; Anbalagan, G.; Pandi, S. Raman and infrared spectra of carbonates of calcite structure. *J. Raman Spectrosc.* **2006**, *37*, 892–899. [CrossRef]
71. Meilunas, R.J.; Bentsen, J.G.; Steinberg, A. Analysis of aged paint binders by FTIR spectroscopy. *Stud. Conserv.* **1990**, *35*, 33. [CrossRef]
72. Mazzeo, R.; Prati, S.; Quaranta, M. Attenuated total reflection micro FTIR characterisation of pigment–binder interaction in reconstructed paint films. *Anal. Bioanal. Chem.* **2008**, *392*, 65–76. [CrossRef]
73. Cerasuolo, A. *Literature and Artistic Practice in Sixteenth-Century Italy*; Brill: Leiden, The Netherlands, 2017; pp. 188–189.

MDPI AG
Grosspeteranlage 5
4052 Basel
Switzerland
Tel.: +41 61 683 77 34

Minerals Editorial Office
E-mail: minerals@mdpi.com
www.mdpi.com/journal/minerals

www.ingramcontent.com/pod-product-compliance
Lightning Source LLC
Chambersburg PA
CBHW061627210926
51850CB00009B/2458
9783725853830